Gifts from your garden

Gifts from your garden

CELIA HADDON

Illustrated with wood engravings by
Yvonne Skargon

MICHAEL JOSEPH : LONDON

First published in Great Britain by Michael Joseph Ltd
44 Bedford Square, London WC1
1985

British Library Cataloguing in Publication Data

Haddon, Celia
  Gifts from your garden.
  1. Horticultural products   2. Gifts
  I. Title
  635   SB318

  ISBN 0-7181-2519-3

Typesetting by Alacrity Phototypesetters,
Banwell Castle, Weston-super-Mare
Printed and bound in Great Britain by
Butler & Tanner Ltd, Frome & London

God Almighty first planted a garden. And, indeed, it is the purest of human pleasures. It is the greatest refreshment to the spirits of man ...

FRANCIS BACON, 1625

For Joelle

Contents

Acknowledgements

These recipes come from a book I have been keeping since I first began to be interested in jam-making and other garden gifts. Most are handwritten – by myself and several others, judging by the scripts. Some are in the form of newspaper cuttings carefully stuck into the book. Therefore I often do not know exactly where they originally came from, though my researches into old cookbooks make me suspect that many of them date back much earlier than their modern authors know. Still, as a gesture towards all those anonymous contributors who have unwittingly supplied me with recipes, I have decided to put aside a proportion of the royalties which I shall eventually give to my favourite charity.

There are plenty of sources I can name, however, including my brother Merlin Haddon, my mother Joyce Haddon, Christopher Dobson, Carol Busch, Peter Barratt, Betty Green, Jack Lang, and I am most grateful to them all. The chief tasters who have helped me have been Ronnie Payne, Betty Mitcham and Dorothy Bradley.

I have also had help and advice from Russell Hobbs and Dobies' Seeds; and I was first shown the beauty of dried flowers by Mrs Elizabeth Bullivant whose home, Stourton House in Wiltshire, is open to the public. Look it up in the National Gardens Scheme book for details.

Together with my publishers, I should like to thank the National Trust and Macmillan London Ltd for permission to quote from 'Our Fathers of Old' by Rudyard Kipling.

Finally, my gratitude to Yvonne Skargon who has made this book look so delightful with her wood-cuts. Readers may like to know that the cat she portrays is called Oscar.

Introduction

Oct. 17 ... Mr. Jeanes sent me this morning a large Hamper of common Apples, I sent the same Hamper back full of my Apples called Beefans with a great many Shrubs, Laurels, etc, etc,.

PARSON WOODFORDE'S DIARY, 1788

Most gardeners are generous by nature. They press cuttings on the visitor, dig up precious plants to give away roots, pluck the earliest and finest of their vegetables, and pick huge bunches of flowers from their herbaceous borders.

These gifts are the natural outpourings of friendship and need no instruction. But some gardeners will want to do more. And the extra work involved in processing the year's harvest – the picking, storing, laying up and preserving – signal the giver's affection the more plainly. It is a joy to hand the visitor a bunch of flowers or vegetables, picked only two minutes earlier. But it is perhaps a rarer delight to give in hot summer the sugared violets which remind the receiver of spring, or to hand over in deepest winter, jam with the summer taste of raspberries.

These garden gifts have always been the privilege of the poor, requiring loving labour rather than a full purse. 'Because I live on a small pension, I cannot afford to spend much upon presents this Christmas. I am making them instead', explained a friend of mine who is a retired nurse. 'I am spending my time rather than my money, and as I do so it seems to me that I am giving something very special.

11

After all, time is part of eternity, so I am giving my family and friends a little bit of eternity in each gift.'

Good gardeners can also feel a proper pride in a gift which shows their skill as well as their generosity, as I do myself when giving away pots of unusual jam, pickles, or sweetmeats made with fruits and vegetables I have grown myself. And the recipes that follow all call for home-grown fruits, vegetables and flowers.

For a gift to give real pleasure and excitement, it should have a certain style to it. It is always nice to receive a pot of home-made jam but for my part I feel that a jar of, say, gooseberry plain and simple without pretty packaging is rather a workaday present. For birthdays, special occasions and Christmas there should be a little more fancy in the gift.

I have often gone back to the old recipe books of Victorian or even earlier England in search of the slightly unusual. In times gone by people were more used to cultivating their own gardens. With no local supermarket nor the wealth of tins, processed foods and exotic imports that we have now, they were forced to make the most of what they could grow themselves.

Considering the provenance of most of the recipes, and my own fondness for anything that is old and traditional, I have decided to stay with the imperial system of weights and measures. But for the more modern-minded, metric tables can be found in the appendix.

I have tried to avoid recipes which involve complicated processes or machinery. Specifically, I have not assumed that everyone now has a food processor: wherever the preparation involves liquidising, I have always suggested an alternative method. In general, I think it a mistake to over-elaborate the simple goodness of the fruits and vegetables themselves. Truly fresh garden produce goes best with simple cooking. Besides this is not a book for the gourmet, but one for ordinary cooks like myself who simply wish to use their gardens to produce pleasure for others, as well as pleasure for themselves.

Equally, I have usually avoided those recipes which can be found in almost any cookbook. Pickled beetroots and onions are, I think, delicious dainties. But there is scarcely any good general cookbook which will not tell you how to make them.

Nor will you find here recipes or suggestions which require very exotic vegetables or fruits. Not all gardeners have the room or the will for such experiments. Instead, I have concentrated on ways of using up the ordinary generous surplus of a good year – recipes for apples, plums, tomatoes and pumpkins – a foison which our forefathers never took for granted.

In their writings there is a kind of wonder at the fruitfulness of their own little patch of earth. 'Gardening and the pleasant delights of a curious Orchard has been the Delight of many great Persons and wise Men of all Ages; who have therein contemplated the Wonder of God in Nature, and refreshed their Minds, as well as sollaced their Bodies' wrote a seventeenth-century gardening writer. And some of the old herbal concoctions and household inventions that he would have recommended are still available to us today, though most have now been lost to modern technology. Of course, an aerospray will scent the wardrobe or the bedroom with unparalleled efficiency as its chemicals permeate everywhere and linger in the atmosphere for weeks. Yet to my mind the more fleeting fragrances of lavender, scented geraniums or dried rose petals make a more delightful gift. Every year I renew the lavender in the dozen or so little bags which lie in my chest of drawers: and every year I remember my mother who made me a round lace-trimmed one, the eleven-year-old niece who sewed the little rectangle with such careful stitches, and Reg who gave me the lavender-stuffed cat.

Two centuries ago in the still rooms of large houses, the maids made up perfumes and sweet waters by distilling. They also made many of the cosmetics which we should now buy over the counter. I have not included still-room recipes of this kind here, since they need equipment which most people will not have.

Also missing are the herbal remedies which can ward off or cure minor ailments. In the ordinary course of things, I do not think it is very heartening to receive medicines as a gift. I have tried to concentrate on those recipes which will give simply pleasure.

Finally, particularly for those without a kitchen garden there are the gifts to be made from an ordinary herbaceous border. In all then, there are gift suggestions for the herb garden, the kitchen garden and the flower border.

Garden gifts will give something of yourself to the recipient in a way that a present bought from the most expensive of shops never could. They will also help those whose purse is stretched to give more generously than they would otherwise be able to. These gifts are not for those you wish to impress but for those you love.

Herbal
Delights

Excellent herbs had our fathers of old –
  Excellent herbs to ease their pain –
Alexanders and Marigold,
  Eyebright, Orris, and Elecampane,
Basil, Rocket, Valerian, Rue
  (Almost singing themselves they run)
Vervain, Dittany, Call-me-to-you –
Cowslip, Melilot, Rose of the Sun.
Anything green that grew out of the mould
  Was an excellent herb to our fathers of old.
RUDYARD KIPLING, 1865-1936

You do not need a special herb garden to make gifts out of herbs. Most are hardy plants which will grow happily in the herbaceous border, or in the kitchen garden among the vegetables, or even in pots on a sunny patio. Most gardens have at least some herbs — mint for new potatoes, parsley, chives and perhaps some thyme grown for its purple flowers.

Herbs should be picked in their prime. 'Of leaves choose only such as are green and full of juice; pick them carefully, and cast away such as are declining for they will putrify the rest.' warned a seventeenth-century herbalist, and his advice holds good today. He also pointed out 'the leaves of such herbs as run up to seed are not so good when they are in flower as before.' The best time to pick herbs is before they are in flower.

I pick mine on a sunny day, after the dew has evaporated, and then dry them simply by placing them on a linen cloth in the spare bedroom out of direct sunlight. Or you can tie up small handfuls with string, and hang them upside down from a beam or picture hook.

Most common herbs dry well like this. Parsley is an exception, turning a pallid white, though it keeps some of its flavour. And basil tastes so much more delicious fresh that for my own kitchen needs I freeze the leaves.

Once the herbs are completely dry, strip the leaves and store them in jars or tightly sealed tins in a dark dry place. The easiest way to make a gift of herbs is simply to put them in similar jars, with a pretty label for each jar.

Seeds, as well as dried leaves, are used for flavouring. Fennel seeds make an excellent sharp-tasting addition to salads or grilled fish. Seeds from the annual opium poppies in the flower border can be used in cakes, biscuits and the famous Jewish delicacy *Hamantaschen*. If you give away little jars of these seeds, write out a recipe for using them and attach it to the jar so that less adventurous cooks know how to use their gift.

A more time-consuming present is home-made bouquets garnis. These little muslin bags containing herbs are designed to flavour food 'invisibly': once the cooking is over, they can be removed before the food goes to table. To make them, either use a sewing machine, or place a saucer face down on some material and cut round it with scissors or, better still, pinking shears. Put the herbs in the middle, gather up the edges of this circle of material and simply tie them together with a piece of thread.

The traditional bouquet garni includes dried thyme, parsley, a crumbled piece of bay leaf, and perhaps a strip of dried orange peel or a very small amount of rosemary. Because the herbs cannot be seen through the muslin, this is one occasion when it is worth drying parsley.

Dried parsley, despite its lack of colour, also plays a part in herb powder. The Victorians used this instead of the assortment of dried leaves we know as mixed herbs, and the bright green powder looks very attractive in jars. It is particularly useful for flavouring barbecue meats.

HERB POWDER

1 oz of dried marjoram, parsley, winter savory, and lemon thyme	4 bay leaves 1 oz dried lemon peel 2 large pinches of celery seeds

The herbs must be completely dry with leaves stripped from the stalks. Place the dried lemon peel in a coffee grinder and grind it up first. Then add the celery seed, and lastly the dried herbs. Keep grinding until they are reduced to powder.

Herbs can be made up into interesting teas – tisanes – for those who are adventurous about what they eat and drink. The various mints are particularly apt – pineapple, orange, eau de cologne or apple mint and, of course, peppermint. Lime tree and camomile flowers and both the petals and leaves of bergamot (*Monarda*) also make pleasant tisanes.

The proper way to prepare them is to pour boiling water on the herbs, and let them infuse for a few minutes. Sweeten with sugar or honey, but do not add milk. Attach these instructions to the jar, if you are giving away a tisane.

For people who might be nervous of drinking herbal teas, try making up these recipes. They both use garden flowers and leaves mixed with Indian or China teas.

CURRANT TEA

24 dried blackcurrant   1 oz good quality
    leaves   Assam tea

Crumble the blackcurrant leaves into small pieces and mix with the tea. Store in an airtight tin.

The delicious flavour of the currant leaves will permeate the tea, yet it is still recognisable as the Englishwoman's 'cuppa'. This can be drunk with milk if required. The following more flowery tea is probably best drunk with a slice of lemon, but once again milk can be added.

LAVENDER TEA

1 oz Lapsang $\frac{1}{2}$ oz dried lavender
    Souchong tea

This combination of black tea leaves and purple lavender looks particularly pretty. Mix together, and put in a tightly sealed tin. Just before sealing the tin, place one dried flower spike of lavender on top of the mixture.

'Mustard is of incomparable effect to quicken and revive the spirits', wrote John Evelyn nearly three hundred years ago. Herbal mustard, combining the bite of the mustard with the flavour of garden herbs, makes an excellent gift. Nowadays it is possible to buy seed-filled French mustard in delicatessens and most supermarkets, though it would be cheaper to make your own version of this delicacy.

1 oz yellow mustard seed

1 oz black mustard seed

1 teaspoon ground ginger

1 teaspoon garlic powder

2 tablespoons dried winter savory

1 teaspoon salt

2 tablespoons olive oil

3 tablespoons water

4 tablespoons wine vinegar

Grind the mustard seeds and dried herbs together. Mix in the other ingredients and leave for 12 hours. Put into a pretty jar.

Mustard seed of both kinds is sold in health food shops, and sometimes in supermarkets. But be careful not to buy the mustard seed which is sold for planting; it may have been treated with poisonous chemicals.

The mixture of black and yellow seed makes a rather prettily speckled mustard. However, those who prefer their mustard sharp may want to use only yellow seed; while those who prefer a mild mustard could use simply black seed on its own.

I think winter savory goes particularly well with mustard, but you could use other herbs. Thyme, marjoram, or rosemary would also taste good. Sage is rather overpowering, I think, for mixing with mustard but this may be a matter of personal preference.

Another condiment easily prepared is herbal vinegar, which brings a delicious new taste to salads and pickles.

chopped herbs   1 fine sprig of the
white wine vinegar   same herb

Chop the herb of your choice into small pieces and add as much to the vinegar as the jar or bottle will hold. Leave the herb to infuse for two to three weeks, shaking the bottle every time you pass it.

Taste the vinegar for flavour when the time is up. Strain the liquid from the herb through a cotton cloth or a jelly bag. Then take a fine sprig of the herb and place it in the flavoured vinegar so that it stretches at least three quarters of the way up the bottle.

Some of the vinegar always gets lost during the straining, so make sure that you start the process with a little more than you think you will need for the bottle you plan to put it in. I would recommend using wine vinegar rather than malt as the latter has such a coarse taste.

Sometimes the chopped herbs simply will not come out of the bottle in which they have been infused. Therefore it is probably wise to make the herbal vinegar in one bottle, and strain it into a second slightly smaller bottle when ready. With some herbs – lovage, for example – the vinegar turns cloudy and even straining won't help. If you leave it for several weeks, these minute particles will fall to the bottom. With a careful hand pour off the clear vinegar leaving the dregs behind.

My own preference is for basil-flavoured vinegar. It is excellent with tomato salad. Vinegar flavoured with elderflowers, which need an extra week's infusion, runs it a close second. The broad-leaved variety of thyme will give a rosy colour which looks attractive. If for some reason the vinegar does not seem very intensely flavoured, simply repeat the process with a second batch of chopped herb.

Another delight is vinegar flavoured with lavender.

LAVENDER VINEGAR

lavender   white wine vinegar

Rub the lavender flowers off the flower spikes. Crush them with pestle and mortar, and fill the bottle of vinegar with as many flowers as possible. Leave for two weeks, shaking daily. Drain off the vinegar and repeat the process.

The vinegar turns a wonderful lavender mauve and seems to lose its tartness in the odour of the flower. Much of the liquid is lost in the straining however, so use treble the quantity of vinegar you want to end up with. It is delicious used to scent a handkerchief or to dab behind the ears.

So strong is the scent that lavender vinegar is almost like a commercial toilet water. The herbalist John Gerard (1545-1612) considered it not just delightful but also medicinal. 'The distilled water of lavender smelt unto, or the temples or forehead bathed therewith, is a refreshing to them that have the catalepsy, a light migram, and to them that have the falling sickness and that use to swoon much.'

Sweet Bags
and
Potpourris

On the backs of my armchairs are thin Liberty silk oblong bags, like miniature saddle-bags, filled with dried Lavender, Sweet Verbena, and Sweet Geranium leaves... The visitor who leans back in his chair wonders from where the sweet scent comes.

A POT POURRI FROM A SURREY GARDEN, 1899

Today we have lost the art of distilling sweet toilet waters from flowers and herbs, or of pressing out their essential oils. The still rooms of the past with their elaborate recipes and equipment are gone for ever. In their place we must content ourselves with trying to capture the delicate scents of the garden in simpler ways.

The commercial soaps and perfumes will always outdo garden recipes in both the strength and endurance of their scents, born of complex chemicals artfully blended in the laboratory and mass-produced in the factory.

The home-made lavender bag or potpourri cannot compete with these. Yet there is something touching in the very impermanence of wrinkled rose petals dried in the summer, or lavender kept in little bags fastened with a ribbon. There is more friendship in these small gifts than in any shop-bought box of perfumes.

LAVENDER BAGS

Pick the lavender on a dry sunny day just before the tiny flowers open. Dry it out of the direct sun on a tray or cotton cloth. The flower heads can then be left intact for several weeks, until the bags are ready for them. Then strip the blue florets off their stalks.

To make the bags, fold the material in two. Sew up the two sides of the rectangle, and hem the edges at the top. Fill with the lavender, then tie the top with a narrow ribbon which matches the colour of the material.

Lavender bags can be of any shape. If you are a good needlewoman, experiment with heart shapes, ovals, circles, diamonds and perhaps even clubs and spades. A modest lace edging – too much lace swamps the bag – will add to their charm.

It is also possible to make animal shapes, adding beads for eyes and strands of bristle for whiskers. Some people even go so far as to make lavender dolls – rag dolls stuffed with lavender and then elaborately dressed. A simpler gift is made with the flower heads intact.

LAVENDER FAGGOTS

Take nine strong stems of lavender fresh cut from the garden. (Dried stems are too brittle.) Tie them tightly together with a long narrow ribbon (about four times the length of the stems) just under the flower head. Then double back the stems and weave one end of the ribbon between them. Nine being an uneven number, the warp of the stems should appear in alternate bands with the ribbon weft to give a pleasing effect. Once the ribbon has completely enclosed the flower heads inside their cage of stems then, taking both ends of the ribbon, make a neat bow to finish off.

Our ancestors would think us rather tame in our choice of fragrances. Today lavender is the only flower we make into sachets. In the past all kinds of herbs were used.

Here is one seventeenth-century recipe for a bag of herbs to induce sleep. 'Take drie Rose leaves, keep them close in a glasse which will keep them sweet, then take powder of Mints, powder of Cloves in a gross powder. Put the same to the Rose leaves, then put all these together into a bag, and take that to bed with you, and it will cause you to sleepe, and it is good to smell unto at other times.'

SWEET BAGS

Dry the leaves of the various mints, sweet woodruff, sweet geranium, marjoram, thyme, and rosemary; dry the petals of scented roses, pinks, lavender and lemon verbena. (For method, see Herbal Delights pp 18-19.) Add one or two cloves, one or two crumbled sticks of cinnamon and a tonka bean. (This last is a traditional ingredient which will have to be bought as you can't grow it yourself.) Experiment with the exact mixture of herbs.

If you are making the bag to help a poor sleeper, then fill it only half-full so that it can lie flat on the corner of the pillow, and add a handful of dried hops out of the hedgerow. Give a baby's nappy pin with your gift, which should be fastened to the pillow in such a way that the sleeper can inhale its fragrance without risking injury from the pin.

These bags can also be made wholly from the sweet geranium. Unlike its gaudy cousins which have bright flowers, this plant usually has insignificant blossoms. But its glory is in the deliciously scented leaves whose strong fragrance can last for months and months.

Like the other pelargoniums, these scented-leaved varieties cannot survive frosts and need wintering on the windowsill or in the greenhouse. Come the summer, plant them out in the flower border where they will grow two or three feet in height once free of their confining pot. In the autumn, having taken cuttings in pots for the next year, I dig up the plants. I usually dry the leaves by hanging the whole plant upside down.

Scented geraniums have delightful names. There is the powerfully scented Prince of Orange smelling as his name suggests; there is Mabel Grey with a citronella scent and Lady Plymouth who smells peppery. For a characteristic scent rather like Turkish delight, try *radula*, attar of roses or *graveolens*.

Peppermint-scented geraniums like Joy Lucille or *tomentosum* may earn a place in bittersweet, rather than sweet bags. *Quercifolium*, also known as the oak-leaved geranium, also belongs in this category.

Herb sachets were originally designed not just to scent clothes, but to keep off moths. As I dislike the smell of mothballs and worse still the evil smell of the new chemical moth killers, I still make up the bittersweet mixture for this purpose. As a gift I think it is an imaginative substitute for the more ordinary lavender bags.

Dry the leaves of wormwood, southernwood, rue, sage, feverfew, oak-leaved geranium and costmary. Add some lavender if the mixture seems to require it. Make a long loop at the top of some of the bags so that they can be hung in the wardrobe; or even use the mixture for stuffing padded clothes hangers.

Of all the bittersweet herbs my favourite is cost-mary, also known as alecost, which has the Latin name of *Chrysanthemum balsamita*. This pungent herb used to be tied up in bundles with lavender and placed in linen presses to keep off moths.

Bittersweet bags filled with a mixture of lavender and costmary leaves are deliciously strong-scented. Costmary is also known as Bibleleaf because a dried leaf was traditionally used as a bookmarker in the family Bible.

The most amusing of all the scented bags is one specially for cats. The grey foliaged plant known as catmint, or *Nepeta x faassenii*, has an odour which attracts male and sometimes female cats. The plant's scent, I have been told, is a chemical compound which is quite close to that of the feline female hormone.

Catmint, in fact, is irresistibly sexy to some susceptible tom cats. 'The smell of it is so pleasant unto them, that they rub themselves upon it and wallow or tumble in it, and also feed on the branches and leaves very greedily' warns the herbalist John Gerard.

Dry the plant as you would any other herb. Leave it intact until you need the leaves. It will keep its odour better like this. Then strip flowers and leaves from the stalk ready for using in the bag, which must be mouse-shaped.

CATMINT MOUSE

Place a saucer face downwards on some material and cut round about a third of its circumference. Then line up the same saucer again, using a third of its circumference to intersect the cut cloth. You should now have an elongated pointed oval. Now blunt one end of it by cutting across the point, rounding it off into a gentle curve. This is the base of the mouse.

Measure along one side of the base and halfway across the curved end. Still using the same material from which you cut the base, fold it double and cut a straight line to the length you have already measured, adding an extra half-inch to allow for hems.

Then cut a curve, with the straight line as base, in the shape of a mouse's humped back. The nose end should be narrow and pointed; the tail end should be high, rounding down quite steeply. Now separate the material, and you will have the two sides of your mouse.

Keeping the material inside out, sew the two sides to the base, then to each other, remembering to attach the humped back to the blunt end of the oval base and the pointed nose to the sharp end. Leave the back end of the mouse unsewn, and turn the material right side out. Stuff it with the dried catmint and sew up all but the final half-inch.

Knot a piece of string or cord at one end and place the knot inside the mouse. Complete the sewing. This makes the mouse's tail.

Use beads for eyes and add two little ears cut out of felt. If you have some bristles, add these by way of whiskers.

A catmint mouse is a light-hearted present. A covered china vase full of potpourri is a rather more serious gift. Today the commercial version is to be found in many shops, wrapped in plastic, usually anointed with mass-produced oils and smelling overpoweringly false.

Take a sniff of these shop-bought petals and your thoughts immediately turn to laboratories. A dried rose smothered in such oils cannot smell as sweet as it should. These ersatz potpourris should be avoided by anybody who loves the natural in life.

To make the real article, you will need one ingredient which must be bought in a shop. It is orrisroot powder. A neutral beige and only slightly scented on its own, it nevertheless brings out the fragrance of other herbs and flowers and, combined with sea salt, acts as a preservative. (The address of a supplier is given on page 119.)

The word 'potpourri' means 'rotten pot', and the most powerful of their kind are just that – a mixture of fragrant petals and leaves which have rotted down together to make a wonderful mingled scent. This is the so-called wet or moist potpourri.

Take the petals of scented roses, especially red ones, and wild roses too, peonies, honey-suckles, wallflowers, pinks, carnations, laven-der, sage, meadowsweet, and *Philadelphus*, known incorrectly as syringa.

Take the leaves of marjoram, lemon thyme, eau de cologne mint, pineapple mint, scented geraniums, bergamot, rosemary, sweet wood-ruff and catmint. Add just a few leaves of costmary, rue, southernwood, wormwood and a few camomile flower heads.

Pick these on a fine day, and place them in layers in a covered earthenware pot. At least half the contents should be made up of rose petals. On each layer scatter orrisroot powder and a generous quantity of crushed sea salt. Layers may be added as the summer passes. They will rot down in the pot.

Once the season is over, seal tightly till just before Christmas. Then add one or two sticks of cinnamon, one or two cloves, and one or two tonka beans, if you have them. Give the contents a good stir with a knitting needle, and transfer to smaller pots.

Moist potpourris should be kept in covered opaque jars, to be opened only when the fragrance is wanted. This way they will last a couple of years or more. If they become too dry, add a little brandy. *Never* use the shop-bought oils which are sometimes sold as potpourri refreshers. They will simply smother the delicious natural perfume.

The other way to make a potpourri is to use dried flowers and leaves. Not so powerfully fragrant as wet ones, they have the advantage of looking prettier and can also be used to stuff sweet bags.

DRY POTPOURRI

Use the same ingredients as for a moist potpourri, but dry the petals and leaves first. Spread them on a cloth out of the direct sunshine for several days. Then add them in layers to a pot, sprinkling orrisroot and sea salt on each layer. Once again, at least half of the contents should be rose petals.

Because a dry potpourri may be kept in bowls where it can be seen, add a few dried petals from marigolds and larkspurs just for their colour. Some dried lemon and orange peel can also be added.

Keep the mixture tightly sealed either in an opaque pot or in a glass jar in a very dark cupboard, till just before Christmas. Mix the contents thoroughly together, and transfer into smaller jars.

Dry potpourris can be given away in transparent containers – old biscuit jars can look pretty. In glass the petals will slowly lose their colour over the period of a year, so it is best to renew them annually.

Potpourri, whether moist or dry, is a gift which has all the scents of the past summer captured for our delight. It is like giving away your whole garden in a little pot.

Sweetmeats
and
Cakes

*Steep Gum-dragant in Rose-water, and soak
your Rosemary-flowers in it: then lay them on a
Paper and strew sifted Sugar over them; lay
them in the hot Sun, turning them, and strewing
Sugar over them till they are sufficiently candied,
then keep them for use.*

THE COOK'S AND CONFECTIONER'S
DICTIONARY, 1723

In the kitchen garden and the orchard grow fruit
and vegetables which can be turned into delicious
sweetmeats and cakes. And that's where the skilled
cook of times past expected to find her ingredients,
fresh from nature rather than the local shops.
Today, spoiled by the freezer and canning industry,
we often forget to use the homely materials which
are merely a minute away down the garden path.

Turning fruits, leaves and flowers into sweetmeats takes a certain amount of skill and sometimes
a great deal of patience. These are true luxuries to be
made in small quantities, and given away to dis-
cerning people who will appreciate the gift. No
harm, though, in dignifying your labour of love with
the right packaging, and I've listed some ideas in the
final section. (See pp 113-117.)

Probably the most familiar way of producing
sweets from the garden is by the crystalising process. There are several possible methods. One is
simply to heat damp sugar to the point at which it
draws out into a thread when a spoon is touched to
the surface. Then dip the flowers or leaves quickly
into the mixture one at a time, dusting them with
caster sugar when they come out.

This method however needs deft movements, and
can too easily result in burned fingers and a mass of

solidified sugar in the pan. An easier way involves a mixture of sugar and egg white.

CRYSTALISED VIOLETS

1 egg white   caster sugar
2 heaped tablespoons   violets
  icing sugar

Whisk the egg white until firm. Add icing sugar slowly, whisking all the time. Using a pair of tweezers, dip each violet in the mixture. Shake off any surplus liquid. Dip a second time if the area covered by the tweezers has not been coated. Shake off surplus.
Strew the caster sugar over a baking sheet, and place the damp violets on top. Dribble more caster sugar on each violet. Put in the oven at its lowest setting, the door left slightly open. Dry them for at least 3 hours.

Exactly the same mixture of icing sugar and egg white can be used to crystalise other flowers and leaves. Try the leaves of apple mint, eau de cologne mint, and balm. Primrose flowers are also particularly pretty when crystalised. For these, and for the leaves, it is probably best to paint on the egg white mixture with a small paintbrush rather than dip them in the concoction.

Try, too, the bright blue flowers of borage. You need first of all to pull away the sepals on the underside of the petals, in order to crystalise only the blue flower itself.

All in all, a fiddly job. Some flowers slip out of sticky fingers. Others break or tear. And you will need to allow for this.

When the sweets are dry, they should be stored in caster sugar in an airtight container. Home-made crystalised flowers are never as neat as the bought – imitation – varieties, but the flavour of the real thing, especially violets, is astonishingly delicious.

Another delightful gift, which must be given away immediately, belongs to strawberry time. If you grow strawberries in your garden, chocolate strawberries will make a fabulous substitute for after-dinner mints.

CHOCOLATE STRAWBERRIES

one dark chocolate   strawberries
   bar

Use the most expensive dark chocolate you can find for this recipe. Melt it in a double boiler, or a pudding basin in a saucepan of simmering water. Using only perfect strawberries with their stalks intact, take each fruit by the stalk and dip it in the chocolate to about the halfway mark. Place carefully on greaseproof paper to dry, if possible keeping the chocolate end uppermost.

The next sweetmeat is, to my mind, the most delicious of all. It needs a great deal of attention and patience in the making, but the result is worth it – something between a fruit jelly and a chewy fudge.

PLUM SWEETMEAT

2 lb plums   granulated sugar
5 fl oz water

Put the water and plums in a saucepan and stew until soft. Take out the stones. Liquidise. Then stir the pulp over a very low heat until it is quite dry. Be careful not to let it burn. Add 1 lb of sugar for each pound of pulp.
Still on a very low heat, continue stirring till the sugar has dissolved and the paste leaves the side of the pan when a spoon is drawn over it. So far, the slow cooking will have taken at least an hour. Turn the sugary pulp into a greased shallow dish, and when it is cold stamp it into shapes.
Transfer to a greased oven dish and place in the oven at its lowest setting. Leave for 2 hours, then turn the plum shapes over and leave to cook for a further 2 hours.
Let them cool, then pack in an airtight container with greaseproof paper between the layers.

Plum sweetmeat will keep for months, though if the container is not airtight it may become rather squashy.

Another strange delicacy which, alas, does not keep so well, is carrot candy. This should be given away shortly after the making. Carrot candy is a traditional Jewish delicacy and I think it makes rather an unusual gift – for gentiles!

CARROT CANDY

8 oz carrots
8 oz caster sugar
2 oz chopped nuts

large pinch ground ginger

Clean and peel carrots. Grate finely. Mix with the sugar in a saucepan, and cook on a low heat until the sugar is dissolved. Continue cooking until the pulp is dry, stirring all the time to prevent it from burning.
Test a little on a saucer. If it hardens as it cools, the mixture is ready. Add the chopped nuts and the ginger and stir them in well.
Spread the mixture in a greased tin, and while cooling mark it into shapes with a knife. When cold and solid, break it along the marks.

Carrot candy is bright orange, with a texture resembling old-fashioned coconut ice. It is not such a treat as plum sweetmeat and does not keep well. But I think it makes an amusing gift, particularly for a gardener who prides himself on his skill with roots.

As well as providing raw materials for sweets, the kitchen garden can supply the ingredients of cakes too. Perhaps the best-known of all is the apple cake. There are many different recipes for this, but my favourite is equally good eaten as a dessert with custard or cream or cut into slices at the tea table. Unlike some recipes this relies on the apple, rather than dried fruit, for its main flavouring and makes a good gift to take with you if you are staying with friends.

APPLE CAKE

2 lb cooking apples
lemon juice or
    vinegar
5 oz butter
8 oz brown sugar

2 large eggs
8 oz self-raising flour
1½ teaspoons baking
    powder

Preheat the oven to Gas Mark 4/350° F/ 180° C. Peel and core the apples and cut them into small pieces about half-an-inch square. Keep these in water with a dash of lemon juice or vinegar so that they do not discolour.
Melt the butter slowly in a saucepan and pour into a mixing bowl. Add almost all the sugar and beat well. Add the eggs, together with the flour which has been sifted with the baking powder.
Now add the apple, spoon the mixture into a buttered cake tin and sprinkle with the reserved sugar. Bake for 1½ hours. For the first 30 minutes place a piece of foil or buttered greaseproof paper over the top of the cake.

I often substitute 100% wholemeal for refined flour, and it is just as delicious. The scrumptious flavour depends largely on using butter rather than margarine, so do not be tempted to economise.

Another garden cake is made from courgettes. This recipe was given to me by an old schoolfriend, Carol Busch, who is married to an American, and it uses American cups as measures. Roughly speaking, these are twice the size of an ordinary English teacup. Don't worry too much about the accuracy of your conversions. The important thing is to use the same system – English or American – throughout.

BESSIE'S ZUCCHINI BREAD

2 small eggs
$\frac{1}{4}$ cup cooking oil
1 cup granulated
  sugar
1 cup peeled and
  grated courgettes
$\frac{1}{4}$ teaspoon baking
  powder
1 cup self-raising flour

1 teaspoon bicarbon-
  ate of soda
good pinch of salt
1 teaspoon ground
  cinnamon
$\frac{1}{2}$ teaspoon vanilla
$\frac{1}{2}$ cup chopped nuts
$\frac{1}{4}$ cup raisins or
  dates

Preheat the oven to Gas Mark 4/350° F/ 180° C. Beat the eggs, oil, sugar and grated courgettes together. Add baking powder, flour, bicarbonate of soda, salt, and ground cinnamon. Mix well. Add vanilla, nuts, and raisins or dates.

Grease a 2-lb rectangular loaf tin and add the mixture which should only half-fill it. Bake for 55 minutes. Let it cool in the tin for 20 minutes before turning out.

This is a nutty cake with a strange but delicious flavour, and it can be eaten with or without butter. I expected a cake made out of courgettes would be rather disgusting. This delicious recipe took me by surprise.

I have tried out the recipe on all kinds of people. Most exclaim with surprise when they learn that it is made with courgettes. So, a good novelty gift from a vegetable which almost always produces a surplus – the answer to a gardener's prayer.

The other vegetable cake is made from carrots. Reasonably familiar to us now as a 'health food', it is delicious as well as good for you. And unlike the other garden cakes, this is a present which can be given in the dark months of winter.

CARROT CAKE

2 oz butter
2 oz dark soft sugar
1 heaped tablespoon honey
2 tablespoons black treacle
4 oz grated carrot
1 large egg
8 oz wholemeal flour
½ teaspoon ground cinnamon

small pinch ground cloves
3 teaspoons baking powder
large pinch ground nutmeg
1 tablespoon demerara sugar
2 tablespoons milk
4 oz sultanas

Preheat the oven to Gas Mark 5/375° F/ 190° C. Put butter, dark soft sugar, honey and treacle in a saucepan over a low heat till they are thoroughly blended together. Pour mixture into a bowl. Add the grated carrot.

Add the egg, and the remaining dry ingredients. Mix together, and add the milk so that the mixture has a dropping consistency. Lastly, add the sultanas.

Put in a buttered cake tin, and sprinkle with demerara sugar. Bake for 55 minutes, or until a skewer inserted in the middle of the cake comes out clean.

This is a dark strongly flavoured cake whose texture is that of a rather heavy fruit cake. It makes a good novelty present at a time of year when courgettes and apples are not available.

Perhaps the best thing about giving cakes or sweetmeats made from garden produce is that they are so unusual. Not only do you show that you care by taking the trouble to cook them, but you will also be giving away a present which will slightly surprise, and surely delight most people.

Cordials
and
Elixirs

To preserve strawberries in wine. Put a quantity of the finest large strawberries in a gooseberry bottle, and strew in three large spoonfuls of fine sugar; fill up with Madeira wine or fine sherry.
THE ART OF COOKERY, 1848

In those far-off innocent days when wines and spirits were virtually untaxed, prudent housewives used alcohol freely to preserve their garden's fruits. With brandy only a few shillings a bottle, all kinds of home-made cordials, elixirs, shrubs and ratafias could be made to hold the flavours and fragrances of summer right into the winter months.

Today, all alcoholic drinks are expensive and thus the country liquors of the past are a costly gift. Yet it seems to me that for special occasions like Christmas, christenings, weddings and birthdays, a small bottle of home-made damson gin, or a jar of fruit preserved in rum is still a worthwhile present, despite the expense.

In France most supermarkets sell special white alcohol, a tasteless but strong spirit in which to preserve fruits. This is rarely, if ever, found at the British wine merchant or off-licence. The best substitute is vodka, a similarly transparent and tasteless spirit though without the strength of the French white alcohol.

The art of home-made cordials is a simple one. In the recipes which follow I have not given exact weights and measures. It is easier to judge proportions by eye and taste. Besides you may simply want to use up an open bottle from the back of the drinks cabinet, rather than buy a new one.

If you decide to use up remains, then remember to allow for wastage in the cooking process. Some will disappear under the guise of tasting; some will be lost in transferring from one receptacle to another. Thus if the final gift is to be a half-bottle, you will probably need to start off with it three-quarters full.

The miniatures of the kind sold on planes and trains can be used to take any overspill from the main bottle. Or, if you cannot afford much of the expensive spirit to begin with, use these little bottles for your gift. A miniature bramble cordial is still a charming present.

Of all the old cordials, the best-known is probably damson gin. Garden cousins of the wild sloes and bullaces found in the hedgerow, these dark rich-flavoured little plums were originally named 'Damascene' for far-off Damascus, an exotic import from the East. Damson gin is a translucent crimson liquor with a fiercely warming taste.

DAMSON GIN

damsons   gin
granulated sugar

Wipe the damsons and take off any stalks. Prick each plum with a darning needle or with the tines of a fine fork.
Fill a jar two-thirds full with fruit, then add enough sugar to come roughly a quarter of the way up the jar. If you plan to give the gin to somebody with a sweet tooth, increase this to a third.
Fill the jar with gin. Seal tightly and leave in a cool dark cupboard for three months, occasionally shaking to help the sugar dissolve. Finally, filter into bottles.

Made in the autumn, damson gin will be ready for Christmas. Do not be tempted to drink it before the time is up, for the full flavour and colour of the damsons will be lost if they do not soak long enough.

Another favourite Victorian drink was a shrub. The word, like 'alcohol' itself, comes from the Arabic and is usually applied to an alcoholic drink based on rum. Shrubs can be made with all kinds of fruit, but the most delicious of them all is based on plums, preferably large juicy red plums like Victorias.

PLUM SHRUB

| plums | granulated sugar |
| lemon peel | dark rum |

Fill jars two-thirds full with plums. Small ones to be pricked all over with a fork; large plums like Victorias are best stoned and halved. Add the kernels to the jar.
Add the zest of one lemon to each large jar. Add sugar a third of the way up each jar, then fill with rum.
Leave in a cool dark place for three months, shaking occasionally to dissolve the sugar. Then filter off the liquid into bottles.

Plum shrub is a delicious drink rather like a liqueur. Do not throw away the rum-soaked plums after straining them out. They can be eaten as they are with cream, making a highly alcoholic but delicious dessert.

One of the oldest English cordials, made with blackcurrants, is now back in favour with the cocktail 'kir'. The drink was invented by a French priest, Canon Kir of Dijon, who mixed blackcurrant cordial with white wine. (Some restaurants serve Kir as a cunning way to disguise rather poor white wine.)

KIR CORDIAL

brandy   blackcurrants
blackcurrant leaves

Bruise the fruit and leaves together and fill a jar two-thirds full. Each jar should contain about two leaves.
Fill the jar with brandy and leave for at least two months in a cool dark place. Then filter the liquid and bottle it.

This recipe makes a sharp-tasting blackcurrant cordial. My husband prefers his rather dry in taste. But for those who like sweet drinks, fill the preserving jar about a quarter-full with sugar before pouring in the brandy. Shake the jar occasionally to help it dissolve.

After straining the liquid, you will be left with brandy-soaked blackcurrants. Sweeten these if you have made the sugar-free version, and eat them in a blackcurrant tart.

One of the delights of late summer are the blackberries that darken the hedges and provide a free harvest for ramblers. Either the wild fruit or its cousins the garden blackberry and loganberry can be used to make this cordial. The recipe comes from the Sussex home of my friend Christopher Dobson.

DOBSON'S BRAMBLE CORDIAL

blackberries or loganberries	split almonds
brandy	cinnamon sticks
	granulated sugar

Fill a jar two-thirds full with the fruit. Add sugar a quarter of the way up the jar. Add two small cinnamon sticks and four split almonds. Keep this in a cool dark cupboard for three months, shaking the jar occasionally. Then filter the liquid into bottles.

Once again, the brandy-soaked berries make a delicious pudding. Add them to some peeled apple, sprinkle with sugar and cover with pastry in a pie dish. Cook as you would an ordinary apple pie. My husband calls this 'The Drinking Man's Pie'.

There is a variant of this cordial which uses whisky in place of brandy. Instead of cinnamon and almonds, place a piece of root ginger and a little lemon zest in each jar. In my opinion, however, the result is not as delicious as when brandy is used.

Of all the cordials I have made the most exotic is one based on sweet geranium leaves. Use only the orange or rose-scented varieties, not the harsh-smelling or peppermint ones.

The geranium, *radula*, has both the right scent, and also leaves which look particularly pretty in the bottle. Its incredible perfume transforms vodka into an elixir which is unlike anything else in the home-made cordial line.

GERANIUM ROSE ELIXIR

scented geranium
   leaves
vodka

one geranium sprig,
   preferably with
   flowers

Snip the leaves and soft stalks into small pieces. Bruise them slightly, using pestle and mortar. Fill a jar, and pour over vodka. Leave the jar in a dark place for a month.

When you come back to it, the liquor will have turned a green-brown colour. Strain off the leaves and add a couple of drops of cochineal or pink colouring. Then bottle the liquor, adding a sprig of geranium for decoration.

This makes a powerfully scented liqueur, so fragant that it is rather like drinking perfume. Those with a sweet tooth may want to add a little sugar to the brew. It is not to everybody's taste, but most people to whom I have offered it have been unable to resist its flowery appeal.

From cordials and elixirs, it is a small step to preserving fruit in alcohol. The most obvious gifts would be apricots floating in apricot brandy, or cherries in cherry brandy – but few gardens grow either. The commoner fruits of the summer, however, make equally delicious presents.

SOZZLED STRAWBERRIES

vodka   strawberries
caster sugar

Choose whole strawberries in good condition. Hull them. For every pound of fruit use just over 5 oz of sugar. Sprinkle on the fruit and leave overnight.
Pack the strawberries in their sugary juice into jars, and fill these up with vodka. Make sure that the fruit is below the surface of the liquid. Keep for at least three months.

To ensure the fruit stays submerged, cover the surface of the vodka with a circle of Porosan (for suppliers, see appendix) before sealing the jar.

52

The strawberries shrink a little in keeping, and the vodka turns a delightful pink. When you come to examine the jar three months later, you may choose to decant the strawberries into something smaller. If so, do not throw away the surplus pink liquid. It is a delicious strawberry liqueur in its own right.

TIPSY GREENGAGES

greengages or yellow plums	granulated sugar vodka

Choose greengages which are firm and not too ripe. Prick with a needle and put in a saucepan, adding cold slightly salted water. Place over a low heat and as each plum comes to the surface, remove it and put in cold salted water. Repeat the process, without letting the water boil. Rinse the greengages. Pack them loosely in a jar and cover with vodka, making sure the fruit is below the surface of the liquor. Leave for two months in a cool dark place.
Strain off the liquor. Make a syrup with 4 oz of sugar and 2 fl oz of water. Add this to the liquor, tasting frequently to make sure it does not become too sweet. Pour it back over the plums and reseal the jars.

This is a rather elaborate process, and it is perfectly possible to preserve plums simply by pricking and sousing them in vodka. But their colour will leech out into the surrounding liquid and though they'll taste perfectly delicious, they won't look as pretty as plums which have been treated the way of the previous recipe.

Another combination of fruit and alcohol is the much homelier one of cheap white wine and apples. This makes a sharp-flavoured dessert, which is delicious with thick cream.

APPLES IN WINE

10 fl oz water	5 lb apples
4 lemons	10 fl oz white wine
2 lb granulated sugar	

Preheat the oven to Gas Mark 2/300° F/ 150° C. Peel the zest from the lemons and add it to the water. Bring slowly to the boil, then leave for 30 minutes. Strain the water into a pan with the juice of the lemons and the sugar. Bring to the boil and simmer until the sugar is dissolved. Add the wine.

Peel, core and slice the apples. Place them in preserving jars – ideally, with a capacity of 1 lb – and fill with the wine syrup. Cook in the oven for 45 to 50 minutes, complete with rubber rings and glass tops if you have the new-fangled jars. If not, you should follow normal procedure for whichever type you own.

Remove from oven and fasten the plastic or metal screwband after wiping any excess syrup from the neck of the jar. Leave for a few minutes then tighten again. When the contents have cooled, check that a vacuum has formed by removing the screwband and carefully lifting the jar by its lid. If the lid comes off, eat the fruit there and then for the bottling process has failed.

Cooks who own a pressure cooker can bottle their apples inside it. Check with a pressure cooker recipe book for details.

These wine-bottled apples have a sharp flavour mainly because of the lemon. Sweet-toothed cooks may halve the amount of lemon and add more sugar to the syrup according to taste.

Remember to attach a pretty label before giving your jars away. Otherwise the recipients may think they have been given some rather ordinary bottled apples, and serve them up for nursery dinner. These apples in wine, thanks to the sharpness of the lemon, have a sophisticated adult flavour, and it is rather a waste to give them to children.

Undoubtedly the most magnificent gift of all using alcohol and fruit is a rumpot (Rumtopf). This is a European invention which has only just reached this country, and it uses lavish quantities of both fruit and rum. To make it, you will need a large and preferably pretty china or earthenware pot with a lid. Nowadays china shops sell special rumpots, but these are often inordinately expensive.

fruits in season   dark rum
caster sugar

For each lb of fruit, use 4 to 8 oz of sugar. Clean the fruit, let it dry, then sprinkle with the sugar and leave for an hour. Pour the fruit and its sugary juices into the pot and cover with rum.

Strawberries, raspberries, loganberries, currants and blackberries can be used as they come. Cherries should be stoned first, as should small plums. Large plums should be stoned and halved. Apples and pears should be peeled and cut into small slices. Gooseberries need cooking gently with the sugar until they are soft; apples and pears can be stewed first too.

To make sure the fruit does not rot, it must be held below the surface of the rum. Use a small plate or a saucer to do this. Continue to add to the pot as the summer proceeds.

The final result is a magnificent rum-laden fruit salad with all the delicacies that your garden has grown. The correct way to eat up a rumpot is simply to put in a large spoon and see what the pot yields. It is a splendidly generous present which could be given to a whole family, rather than just to an individual.

Jams
and
Jellies

We have rose-candy, we have spikenard,
Mastic and terebinth and oil and spice,
And such sweet jams meticulously jarred
As God's Own Prophet eats in Paradise.
                    JAMES ELROY FLECKER, 1884-1915

How charming it is to think of Mohammed in some Eastern paradise, surrounded by houris offering little jars of jam – home-made, I hope, by some celestial Women's Institute. For if there is such a thing as virtuous temptation, then for me it is the sight of the Women's Institute stall at a local fete loaded with home-made jam – a substance which bears no resemblance to the wodge of unreal colour that comes out of most bought jars.

Having started to make my own, I found it almost impossible to enjoy bought jam ever again, especially once I knew how easy a delicacy it is to prepare – easy with the fruits that are rich in pectin, the ingredient which makes jam set, such as goose-berries, blackcurrants, redcurrants and plums. Strawberries, raspberries and blackberries, among others, do not set so easily and will usually need extra lemon to help out.

Recipes for straightforward, single-fruit, jams like these can be found in most recipe books. Those that I give away as presents, rather than keep for my own use, tend to be slightly more unusual. They are jams that you are unlikely to find on supermarket shelves.

Each recipe will make approximately five 1-lb jars unless otherwise stated. It is a good idea to save the tiny jars that are given with breakfast on trains and planes, and use these to take any surplus. At the end of the summer you will find you have half a dozen of these filled with different jams – a very acceptable present in themselves.

The principles of jam-making are simple. The fruit must be cooked to release the pectin *before* the sugar is added. Sugar toughens fruit skins. Then, after the sugar has dissolved, the mixture is cooked at a high temperature.

The easiest way to ensure that the jam will set is simply to test it. Put half a teaspoon of the mixture on a cold plate. Leave it for two or three minutes in the fridge, then push at it with a finger. If the surface wrinkles, the jam is ready to pot.

All kinds of garden produce, vegetables as well as fruit, can be made into jam. The common jam of the thrifty Victorians was rhubarb. My father still shudders at the thought of it, since it was served up day after day by his Nanny in the nursery as a kind of penance. But today, when rhubarb jam is no longer common, I consider it well worth a place at the drawing-room tea table.

Because rhubarb will not set on its own, it needs the addition of either lemon or orange. My favourite recipe is an Edwardian one for a jam which would also suit the breakfast table.

RHUBARB MARMALADE

1 pint of rhubarb cut into small pieces	3 small oranges
	1 lb granulated sugar

Using a sharp potato peeler, strip the zest from the oranges. Cut it into tiny strips. Remove the pith from the fruit itself, and cut the flesh into slices.
Put the rhubarb and the oranges together with their peel into a saucepan, and simmer till the

rhubarb is soft. Add the sugar and cook gently till it is dissolved. Boil rapidly at a high temperature till setting point.

Those who do not have a pint container can make this recipe by using just over 8 oz of rhubarb. These quantities make only 2½ lb of jam. As this is an unusual jam, it is probably worth trying first with only small quantities. If you enjoy the taste, double the quantities next time.

Rhubarb can also be used in the next recipe – as can marrow, courgette or pumpkin. Marrows make a pale yellow jam, courgettes a lime-green and pumpkins a bright golden one. I am told that if you make this recipe with one of those harvest festival pumpkins of enormous size, the jam may turn a rosy pink. Rhubarb makes an agreeable pale red jam.

If you use a marrow, or a fully grown pumpkin for this recipe, scoop out the seeds and pith. With courgettes and very young pumpkins, these can be left in the jam giving a pleasing nutty taste and texture. If in doubt, test one of the seeds by biting it. If it is completely soft, you can use the seeds in the jam.

PUMPKIN, COURGETTE OR MARROW JAM

4 lb of your chosen vegetable peeled (this excludes the weight of pith and seeds if too tough to be used)	4 lb granulated sugar 1 oz root ginger 4 or 5 large lemons

Cut the peeled pumpkin, courgette or marrow into half-inch cubes. Place these in a large basin and sprinkle with 1 lb of the sugar. Leave overnight. The next day bruise the ginger with a rolling pin and tie a small piece of string round it, so that it can be easily retrieved from the jam mixture.

Grate the zest from 2 or 3 of the lemons, and squeeze out the juice from all 5. Place all the ingredients, except for the 3 lb of sugar remaining, in a large saucepan and simmer gently for about 30 minutes till the marrow looks transparent.

Add the rest of the sugar, and boil rapidly. Test for setting. Before potting, remove the ginger root.

This jam may need boiling for some time before it will set. Do not panic. Just keep boiling it, stirring it to make sure it does not burn. The recipe makes about 6 lb of jam.

Gooseberries are rather a glut in my garden, and I have to pick off the unripe berries early to thin out the harvest. With these, I make gooseberry curd.

GOOSEBERRY CURD

3 lb green goose-	$1\frac{1}{2}$ lb caster sugar
berries	4 oz butter
15 fl oz water	4 eggs

Top and tail the gooseberries. Cook in the water till soft. Pass the fruit through a sieve, or liquidise it thoroughly. Sieving gives the better texture.

Put the pulp into a double saucepan or *bain-marie*. Add the sugar and the butter and stir till dissolved. Beat the eggs thoroughly, then add them to the mixture. Cook gently until the mixture thickens.

Unlike most jams, gooseberry curd does not keep. An unopened jar will last for about a month in the refrigerator; an opened jar has a life of only two weeks. (Gooseberry curd can be frozen, however, in plastic pots.)

This means that gooseberry curd is a gift which must be given away as soon as it is made. Make sure the recipient knows that it must be eaten instantly.

GOOSEBERRY CHEESE

3 lb gooseberries   granulated sugar
10 fl oz water

Top and tail the gooseberries. Simmer them in the water until soft. As with the curd, sieve the fruit – liquidising does not give a smooth enough texture.

Allow 12 oz of sugar for each lb of pulp and stir it in. Cook gently till it is dissolved and continue cooking on a very low heat.

The cheese is ready when a spoon drawn across the bottom of the pan leaves a clear line. This will mean about an hour's cooking, possibly more if the fruit is juicy. Stir constantly to make sure it does not burn.

Gooseberry cheese should not be put into jam jars, but a receptacle with a wide neck – a small pudding basin, for instance, covered with greaseproof paper. The cheese is turned out as a kind of solid jelly – to be eaten in slices with cream or thick yoghurt.

Make sure you label your gooseberry cheese, and explain how it is used. Otherwise the recipient may not realise what delicacy they have been given.

Some of the most delicious jams come from a mixture of fruit. Combining two together has the advantage of helping the jam to set. The next recipe also scores in making strawberries go a little further, by adding the more prolific gooseberries. And in my opinion, the tart flavour added by gooseberries actually makes this mixed jam superior to straight-forward strawberry jam.

STRAWBERRY AND GOOSEBERRY JAM

1½ lb gooseberries   1½ lb strawberries
5 fl oz water   3 lb granulated sugar

Top and tail the gooseberries. Simmer them in water until soft.
Hull and mash the strawberries. Add them to the pan and simmer for a further two minutes.
Add the sugar and cook gently till dissolved.
Boil rapidly until setting point.

If you are short of strawberries, you can alter the proportions of fruit so that it is two thirds gooseberries and only one third strawberries. In the same way, the jam can be made with two thirds strawberries and one third gooseberries. But do not reduce the gooseberries any further, otherwise the jam may not set.

Redcurrant jelly has a flavour I do not much enjoy, but a glowing colour which is incomparably beautiful. To get the colour without the flavour, I mix redcurrants with raspberries which benefit from the former's high pectin content.

RASPBERRY AND REDCURRANT JAM

1 lb redcurrants	2 lb raspberries
5 fl oz water	3 lb granulated sugar

Cook the redcurrants in the water gently until they are soft. Add the raspberries and simmer for another 5 to 10 minutes. Add the sugar, and continue to simmer until it dissolves. Boil rapidly until setting point.

This delicious jam has a slightly gritty texture from the redcurrant, a texture which I enjoy. However if you are going to give it to an elderly person with false teeth, it might be better to strain off the redcurrants in a sieve, using only the juice.

Later in the summer comes the bountiful harvest of plums, apples and pears. In a good year, just one tree can produce pounds and pounds of fruit. My neighbour has a Victoria plum that provides jam, chutney, frozen stewed fruit and fresh ripe plums for three or more households every September.

The following recipe uses plums and pears together. It is best made with damsons or the rather old-fashioned dark plums. Greengages and the very sweet green plums make too sickly a jam.

PLUM AND PEAR JAM

| 2 lb ripe pears | 10 fl oz water |
| 2 lb plums | 4 lb granulated sugar |

Peel and core the pears and cut into half-inch pieces. If the pears are hard, then stew them in the water until soft before adding the rest of the ingredients. Otherwise, put plums and pears together and simmer in the water till soft, removing the plum stones as they rise. Stir in the sugar and cook gently till it is dissolved. Boil rapidly until setting point.

The pears give this jam a seductive taste and texture. If your plums and pears do not ripen at the same time, freeze one until the other is available.

Plums also make a Victorian favourite with the charming name of plum gumbo. This has something of the taste of marmalade and is a truly delicious breakfast jam on toast.

PLUM GUMBO

4 lb plums
2 small oranges
1 lemon

8 oz seedless raisins
3 lb dark moist sugar
4 oz split almonds

Stone the plums. Finely grate the zest from the oranges and lemon. Squeeze out the juice. Put zest, juice, raisins and plums into a saucepan and cook gently.

Liquidise the mixture. Add the sugar and cook gently till it is dissolved. Add the nuts, and boil rapidly until setting point.

The latest of all the fruits to ripen is the quince. 'There is no fruit growing in the land that is of so many excellent uses as this, serving as well to make many dishes of meat for the table, as for banquets and much more for their physical virtues', wrote the seventeenth-century herbalist John Parkinson.

There is something magical about quinces. Just one golden fruit in the kitchen scents the air, and a little slice or two of quince in an apple pie metamorphoses that homely dish into something rich and strange.

Quinces are rarely sold in greengrocers. If you are lucky enough to have a quince tree in your garden, or even a japonica – full name, *Chaenomeles japonica* – which produces fruit, you will be able to make the queen of all jellies.

QUINCE JELLY

2 lb quinces	2 teaspoons cream of
3 pints water	tartar, or juice of
granulated sugar	2 lemons

Cut the fruit into pieces no larger than two-inch cubes. Simmer with water and cream of tartar, or lemon juice. Cook until tender, probably about one hour.

Strain through a jelly bag. Allow 1 lb of sugar to each pint of juice. Dissolve the sugar in the juice and boil rapidly till setting point.

A second extract can be made from the same quinces, using half the quantity of water and simmering for 30 minutes only.

Most quinces are pear-shaped fruits, but there are some like apples. I have come across one variety which is round, yellow and very sour. If this is the kind of quince you have, you will not need to add cream of tartar or lemon juice.

The more usual pear-shaped quinces make a glowing rose jelly, full of fragrance. They can also be made into a kind of marmalade with apples. John Gerard wrote enthusiastically that 'The Marmalad or Cotiniat made of quinces and sugar is profitable to strengthen the stomach, that it may retain and keep the meat therein untill it be perfectly digested.'

QUINCE AND APPLE MARMALADE

2 lb quinces	2 level teaspoons
3 pints water	cream of tartar or
apples	the juice of 2
granulated sugar	lemons

Cut up the quinces and simmer in the water with the cream of tartar or lemon juice, just as you would for quince jelly. Strain through a sieve, or a colander would do for in this recipe the odd fragment of quince in the liquid does not matter.

Allow a pound of cut and peeled apples to every pint of liquid. Boil these together till the apples are soft. Add 1 lb of sugar to each pint of fruit pulp, and cook gently till the sugar dissolves. Then boil rapidly till setting point.

Quince and apple marmalade, like quince jelly, makes a first-class gift. It is not a jam you can buy from any normal grocer. It is a jam that many people have never even had the chance of trying. Beg or buy quinces from your neighbours, if you have none yourself; or plant your own quince tree.

I believe that the Prophet himself would not say 'No' to a jar of quince jam. To my mind, it is a positively paradisiacal gift.

Pickles
and
Spiced Fruits

On a hot day in Virginia I know of nothing more comforting than a fine spiced pickle brought up trout-like from the sparkling depths of that aromatic jar below the stairs in Aunt Sally's cellar.

THOMAS JEFFERSON, 1743-1826

Pickles are a way of preserving fruit and vegetables with the sourness of vinegar, just as jams preserve them with the sweetness of sugar. The traditional British pickle, however, is a fierce concoction of malt vinegar which to my mind smothers rather than enhances the taste of the contents.

In the recipes that follow I always use wine vinegar, because it seems to me to make the vastly superior pickle. For spiced fruits, it is essential and I have specified as much. Elsewhere I have merely mentioned vinegar, and those with a passion for the malt variety may use it if they choose.

Unlike chutneys and ketchups which mix garden produce together, pickles should preserve both taste and texture of the individual fruit or vegetable. Sometimes the contents are cooked a little beforehand to let the juices run. Some pickles are fiercely hot, others are sweet and may set like a tremblingly weak jelly.

As most cookbooks have recipes for pickled onions, beetroot, cucumbers and piccalilli, I shall not repeat them here. Like the more common jams, they make tasty gifts but lack something in the way of imagination. Slightly unusual pickles are more fun to give away.

Not all the recipes specify quantities. When I am making pickles I often set out to fill three jars, rather than to pickle 4 lb of fruit. Therefore I pick

70

enough fruit or vegetables to fill the jars, and a goodly amount over to allow for shrinkage in cooking. I decide how much vinegar to use on the same basis, that is, whatever is needed to fill the jars. Besides, some of my favourite recipes never did have quantities in the first place.

In most gardens, the first fruit to be truly plentiful is the gooseberry. 'The fruit is much used in dinner, sauces for meats and used in broth instead of verjuice,' said John Gerard.

Once you have eaten gooseberries in tarts, pies, and summer puddings, made jam with them, frozen them, why not pickle them? Their tart flavour makes a fine pickle for eating with rather fatty meats such as pork or ham.

The gourmet version of this recipe would be made with elderflower vinegar. If this is not conveniently to hand, add two or three elderflowers to the vinegar while cooking, taking them out before you seal down the pickle.

GOOSEBERRY AND ELDERFLOWER PICKLE

$1\frac{1}{2}$ lb granulated sugar
1 pint vinegar

1 teaspoon mixed spice
$2\frac{1}{2}$ lb gooseberries

Melt the sugar in the vinegar over a low heat, adding the spice. When the sugar is dissolved, add the gooseberries and simmer. *Before* they begin to burst, take them out with a slotted spoon and put in a jar.
Boil the vinegar rapidly until it is reduced. Pour over the gooseberries.

71

This quantity fills two 7 oz instant coffee jars. For more, just multiply the quantities accordingly.

Few gardens grow their own cherries, and I have therefore avoided most of the recipes which use this fruit. Nevertheless I cannot resist this one, given to me by my brother Merlin. It is best with the sour Morello cherries he has in his French garden, but will work with sweet cherries too.

MERLIN'S PICKLED CHERRIES

granulated sugar   cherries
red wine vinegar

Allow 4 oz sugar to each pint of wine vinegar. Boil the sugar in the vinegar until it is dissolved. Let the liquid cool. Remove stalks and stones from the cherries. Place them in a jar and pour over the cold vinegar.

This makes a sophisticated pickle that is particularly nice with roast duck. You can spoon some of the cherries and vinegar into the roasting pan and incorporate them with the gravy while the bird is cooking; or, serve them with the meat cold. This fruit pickle is less sweet than most of its fellows.

Later in the summer comes the glut of plums. The Victorians, who could not rely on imported foods like we do, used to make them into a kind of substitute for olives. This is a particularly useful recipe since it can use up the small green plums which are blown off the tree before harvest time.

white wine vinegar
pinch of salt
green plums

1 teaspoon mustard
seed for each pint
vinegar

Simmer the vinegar with salt and mustard seed.
Put in the green plums while mixture is still
hot. Simmer for one minute. Allow to cool,
then pot.

I have not specified quantities for this recipe, since
much will depend on the plums available. If yours is
a young plum tree, you may not want to sacrifice too
many of its fruits at this early stage.

An unusual pickle, it is not one to make in bulk
anyway. But a small jar makes a nice present. They
do have a slight echo of the taste of olives, and there
is also a hint of pickled walnuts. They certainly look
like the real thing but, of course, do not have the
genuine texture.

As the plums ripen, they call for different treat-
ment. Damsons and the other small dark old-
fashioned varieties make the best pickle, but the
large modern plums like Victoria can also be used.

DAMSON OR PLUM PICKLE

6 strips of lemon peel
juice of one lemon
2 blades of mace
10 fl oz vinegar

1 lb granulated sugar
2 lb damsons or plums
10 blackcurrant
leaves

Put lemon peel and juice, mace, vinegar and
sugar into a saucepan and cook gently till the
sugar has dissolved. Add the fruit. (If you are

using large plums, these should be pricked all over with fork.) Bring the mixture back to the boil.

Simmer the fruit until their skins begin to break. Take them out with a slotted spoon and put in jars. Divide the mace, lemon peel and blackcurrant leaves between the jars.

Boil the vinegar to reduce its volume and pour over the fruit while still hot.

Both plums and damsons lose some of their volume in the cooking – exactly how much will depend on the variety and the cooking time. But for damsons, it is as well to allow $1\frac{1}{4}$ jars of uncooked fruit for every jar of the cooked damsons.

An even simpler pickle is made from apples, onions and green chillies. This is another very old recipe, with a name which suggests to me that it may have been thought an aphrodisiac. It certainly makes a sharp and fiery pickle.

LADIES' DELIGHT PICKLE

10 fl oz white wine vinegar
1 teaspoon salt
4 oz onions

4 oz apples pared and peeled
1 oz green chillies

Boil the vinegar with the salt and leave to cool. Either chop or mince together the onions, apples and chillies. I sometimes liquidise them. Simply put the mixture in a jar and pour over the cold vinegar.

These quantities fill one 7 oz instant coffee jar – probably enough since this is such a fierce pickle it is not needed in large quantities.

The delicious thing about Ladies' Delight is its freshness. Although it is gaspingly hot, you can taste the apple and onion through the heat – especially if the apple is left unpeeled.

At the onset of autumn the hedgerows start offering their tempting harvest of free blackberries. In good years it is difficult to know what to do with the enormous quantities of fruit available. Pickling is one solution.

BLACKBERRY PICKLE

2 lb ripe blackberries   1 oz allspice
½ oz ground ginger   10 fl oz vinegar
1 lb granulated sugar

Mix the blackberries with the ginger and leave for 12 hours. Then dissolve the sugar and allspice in the vinegar over a low heat. Add the blackberry mixture and simmer till the fruit turns red. This should take about five minutes. Pack the blackberries into a jar. (They will have shrunk to half their volume during cooking.) Boil the vinegar rapidly until reduced by half, then pour over the blackberries.

A slightly milder version of this pickle can be made using ground cloves and ground cinnamon in place of the ginger and allspice. But I prefer the bite given by the ginger.

While pickles can be hot and fiery, spiced fruits are always mild. According to taste, they can be eaten with cold meat, put round a roast joint, or eaten with cheese.

Spiced fruits are preserved by a mixture of vinegar and sugar. To make sure that they do not go bad, heat the jars in an oven set to Gas Mark 2/300° F/ 150° C before filling them.

SPICED APPLES

6 tablespoons honey
8-10 eating apples

5 fl oz white wine
vinegar

Put honey and vinegar into a saucepan and heat gently until the honey is dissolved. Peel and core the apples and cut into large slices – 8 per apple, or 12 for particularly large apples. Add the slices to the pan and simmer gently for 10 minutes, until the slices are almost transparent but before they begin to break up. Spoon them into hot jars and pour over any extra vinegar.

This makes about enough spiced apples to fill a 2 lb preserving jar. They taste delicious, but do not use until at least two months have passed, although they will keep for almost as long as you like. It is a recipe I find very useful as I have a tree which produces large quantities of eating apples that do not store well. I preserve them spiced in this way.

Finally, a recipe for spiced quinces. If you only have a few quinces, use them to make jelly and pep up apple dishes. But should you be lucky enough to have a glut, these spiced fruits will make a charming gift.

SPICED QUINCES

quinces
water
pinch of salt

granulated sugar
white wine vinegar

Peel and core the quinces. (Reserve peelings for quince jelly.) Cut each fruit into eight pieces and cover with cold water, adding a pinch of salt. Bring to the boil, and simmer for ten minutes. Strain off the fruit. To each pint of remaining liquid add 1 lb sugar and 5 fl oz white wine vinegar. Bring to the boil and add the quinces. Simmer until the fruit is tender. Leave for 24 hours.

Drain off the quinces and boil the liquid rapidly to reduce its volume, for up to half an hour if necessary. It should be a glowing pink if you have used the pear-shaped variety. Pack the fruit into hot preserving jars, pour over the liquid and seal.

Spiced fruits are rather an acquired taste, except perhaps for the apples whose piquant flavour as an accompaniment to cold meats will be familiar from apple sauce. Certainly most people will not have come across these delicacies before, so label clearly – especially the spiced quinces which will probably be mistaken for bottled fruit. Make sure the person you are giving these to knows that they are eaten with savoury dishes.

It is possible to eat spiced quinces as a dessert with thick cream. I have done so. But they are rather strong-tasting for a sweet course. I prefer them as a garnish for a meal of cold meats.

For fun, if you are making more than one pickle during the summer, you could fill an extra small jar on each pickling occasion. Then you will have a variety, all in matching little jars – a delightful present.

Chutneys
and
Ketchups

Good home-made ketchup is a valuable addition
to the storeroom, and a good housekeeper will
always look with pride upon it as it stands in
closely corked bottles, neatly labelled, upon her
shelves, feeling as she may that she possesses close
at hand the means of imparting a delicious
flavour to her sauces and gravies ...

CASSELL'S *DICTIONARY OF COOKERY*, 1905

Both chutneys and ketchups are a testimony to the
enterprise of the British merchants who set up shop
far afield along the shores of what was to become the
British empire. Chutney comes from India, the
original word being the Hindi '*chatni*'. Ketchup
comes from China, from '*ke-tsiap*', meaning the
brine in which fish has been pickled. Both words are
now wholly naturalised as part of the English
country cook's vocabulary.

Chutneys, unlike pickles, do not preserve the
original texture and flavour of fruit and vegetables.
Instead a mixture of the two is simmered to a jammy
texture, the thickness and stickiness of which vary
with the recipe. Indeed one may think of it as a kind
of savoury jam. I once made a rather pleasant apple
chutney which my father ate by mistake on bread
and butter for his tea. His only comment: 'How did
you get this jam so sour, Celia?'

The great temptation facing chutney-makers is
simply to throw in whatever surplus fruit and
vegetables are to hand, adding random handfuls of
dried fruit from the larder. This use-it-up chutney is
usually better than the commercial article and can
sometimes be spectacularly good.

For gifts, a more careful selection of contents
is worth the extra effort. If a-bit-of-everything

chutney is delicious, then a carefully thought-out chutney is the gourmet version. I have tried, however, to stick to recipes using fruit and vegetables that normally grow in abundance.

Everybody with an apple tree knows the mingled pleasure and pain of a good apple year. On the one hand there is the joy of picking the abundant harvest; on the other there is the guilt brought on by seeing apples rotting in the grass, scorned even by the sated blackbirds.

There are plenty of recipes which use up apples. This is one of my favourites. The addition of lemon makes for a sharp-tasting chutney which nevertheless keeps its apple flavour. Just as I use wine vinegar in all my pickles, so I would recommend it for all ketchups and chutneys. I consider it worth the extra few pence, but those who truly prefer malt vinegar can use it in any of the recipes in this chapter.

APPLE AND LEMON CHUTNEY

3 lb cooking apples	1 teaspoon salt
1 lb onions	ground pepper
5 fl oz water	1 pint vinegar
1 tablespoon ground cinnamon	juice and grated zest of 3 lemons
1 tablespoon ground ginger	1½ lb moist brown sugar

Core and peel the apples. Peel and chop the onions. Simmer these together for about 25 minutes with the water. Add the spices, salt, pepper to taste and half the vinegar, together with the lemon juice and the grated zest. Cook till the apple is soft. Then add sugar and the rest of the vinegar. Simmer till the mixture is smooth and thick.

Because I am fond of fibre, I sometimes leave in the apple peel. This will mean the chutney is not quite as smooth – an effect I sometimes deliberately exaggerate by undercooking the apple so that though most of it is soft some pieces remain separate. I rather enjoy coming across chunks of apple in my chutney.

Another way to use up apples is to combine them with plums and pears to make orchard chutney. If your fruit does not ripen together, freeze the first comers to await the later ripeners. If you are using slightly unripe pears, you will have to stew them in a little water until they soften before beginning this recipe.

ORCHARD CHUTNEY

8 oz onions	cloves
2 lb of apples, plums and pears mixed	good pinch of salt and pepper
1 teaspoon ground ginger	1 teaspoon mustard seeds
1 teaspoon mixed spice	13 fl oz vinegar
	4 oz chopped dates
1 teaspoon ground	6 oz demerara sugar

Peel and chop onions finely. Core and peel apples and pears. Stone plums. Boil the seasonings in vinegar for 15 minutes. Strain off the mustard seeds, and add the spiced vinegar to the onions, fruit and dates and cook till the fruit is soft. Then stir in the sugar and simmer till the mixture has a jam-like consistency.

The other great burden on the successful gardener is a pumpkin. Pumpkins are irresistible to grow, but difficult to use up once you have decided to cut short the life of the yellow monster. When I finally decide to harvest a pumpkin I have been growing all summer, I eat it as a vegetable, I make soup of it, I make jam, and then in increasing desperation I finish it off as chutney.

If your pumpkin is a dinosaur of great age, then make sure you take out all the seeds and the cottony centre before cooking it. If you have plucked it at a more tender age and its seeds are still soft, you can leave these in the chutney.

PUMPKIN CHUTNEY

$2\frac{1}{2}$ lb pumpkin
8 oz onion
2 cloves garlic
1 lb cooking apples
2 oz sultanas
1 pint vinegar
1 teaspoon ground allspice
good pinch of salt and pepper
12 oz moist brown sugar

Peel the pumpkin, and take out seeds and pith. Cut it into small pieces – one-inch cubes, roughly. Peel apples and slice. Peel and chop onions. Slice or crush garlic cloves. Put all the ingredients into a saucepan except for the sugar and cook gently till the apple and the pumpkin are soft. Add the sugar and continue cooking until the chutney is thick. This may take some time if it is a very watery pumpkin.

This chutney is a relatively unsticky version of a recipe that can be used on surplus marrows or courgettes as well. Don't bother to core courgettes, or even to peel them if they are very young.

The next recipe can be made earlier in the year than most chutneys since it uses rhubarb. It is odd to think what a latecomer rhubarb is to our gardens. In 1732 the first seeds came to the West from a Tartar rhubarb merchant, and apothecaries in Britain began selling it as a medicine. It was not till the beginning of the nineteenth century that the stems were sold fresh to eat just for pleasure.

If you decide to make this relatively late in the summer, discard at least two inches of the stalk together with the leaf.

RHUBARB CHUTNEY

10 fl oz vinegar
5 fl oz water
8 oz demerara sugar
½ teaspoon whole cloves
1 teaspoon mixed spice
good pinch salt
½ teaspoon ground mustard
1 teaspoon celery seeds
2 onions
1 lb rhubarb
8 oz seedless raisins

Simmer vinegar, water, sugar and seasonings together in pan for five minutes. Chop the onions, and cut rhubarb into small pieces. Add these to the liquid and simmer gently for 30 minutes. Add the raisins and simmer for a further 15 minutes. Then continue cooking till the mixture thickens, stirring to make sure it does not burn.

This makes three 1 lb jars of chutney, tasting quite strongly of the raisins whose flavour seems to mix well with the tartness of rhubarb. Those who like their chutney very sweet may want to add a bit more sugar to the recipe.

If chutneys have much in common with jams, ketchups are a far cry from either. They have a very different consistency – definitely more liquid – are meant to be used as sauces, and usually have less complex mixtures of ingredients.

It is essential to find wide-mouthed bottles for ketchups, otherwise the recipient will be left with the tantalising smell of a sauce which refuses to emerge. If in doubt, put them in jam or preserving jars so that they can at least be spooned out.

After filling the containers, I sterilise my ketchups in a pressure cooker, standing the bottles on the platform and wrapping each in a piece of newspaper so they do not clank together. Leave the tops on loosely. I heat mine to a low pressure of 5 lb, giving them only two minutes. This process means that the ketchup will keep for longer.

The other method of sterilisation is to put the bottles in an ordinary pan, wrapped and standing on paper. Immerse them up to their necks in water and simmer for 20 minutes. When either sterilisation process is over, tighten the lids immediately.

TOMATO AND BASIL KETCHUP

3 lb ripe tomatoes	2 large sprigs of basil,
2 cooking apples	fresh or frozen
1 medium onion	4 oz granulated sugar
good pinch salt	13 fl oz vinegar
good pinch paprika	

Peel the tomatoes after placing them for a minute in boiling water. Peel and core the apples. Chop onion. Simmer tomatoes, onion, apple, salt and paprika in a pan with the lid on. When it is thoroughly cooked, add the chopped basil, leaving out any tough stalks. Then liquidise thoroughly so that no chunks remain.

Return the mixture to the pan, add the sugar and vinegar and cook uncovered till it acquires a thickish consistency, something like lemon curd. Bottle and sterilise.

This makes seven or eight bottles about the size of a small Lucozade bottle. It is not worth using anything much larger.

Hot Apple Ketchup is for people who like curries and peppery foods. It looks deceptively mild, like apple sauce, but has quite a bite. Increase the number of chillies and cardamoms if you're cooking this for a real curry fiend.

HOT APPLE KETCHUP

2 lb cooking apples
1 lb onions
2 green-chillies
8 green cardamoms
1 pint vinegar

2 teaspoons ground ginger
6 cloves garlic
6 oz granulated sugar

Peel the apples and onions and chop them. Chop the chillies. Stew these with the spices, vinegar and garlic until the apple is soft. Add the sugar and simmer gently until it is dissolved. Fish out the cardamoms, and then liquidise all that remains very thoroughly. If you want a smoother texture, sieve it. Reheat to boiling. Bottle and sterilise.

Finally, a leaf out of a cookery book first published in 1856. The recipe comes from Bengal, and relies on the use of very acid fruit. I make this early in the year when the cooking apples are still hard, green and completely unripe. It can apparently be made with sloes or even crab apples, though I cannot vouch for this from experience.

BENGAL CHETNEY SAUCE

8 oz sour green apples
4 oz seedless raisins
3 cloves garlic
4 oz moist brown
  sugar

2 oz ground ginger
2 oz cayenne pepper
2 oz salt
vinegar

Peel and chop the apples and liquidise them with the raisins and garlic cloves. Add sugar, ginger, pepper and salt. Beat everything together till well-blended. Add vinegar until the mixture is the consistency of thick cream. Bottle and cork tightly.

'This sauce keeps better if exposed to a gentle degree of heat for a week or two, either by the heat of the fire or in a very full southern aspect of the sun' advises the Victorian book. I placed my two small bottles on the windowsill for three weeks.

To my surprise the sauce does keep well, presumably because the vinegar preserves it. It has a very fierce sharp taste which will be enjoyed by those who like curries. It also uses up those hard green windfalls which are too sour even for the birds.

Cold meats are rather monotonous meals which need cheering up with chutneys and ketchups. But variety is the key, not quantity. Do not be afraid to fill small containers, for a little goes a long way. And as I suggested in the chapter about pickles and spiced fruits, three small jars or bottles containing three different condiments make a much nicer present than one huge one on its own.

Everlasting
and
Dried Flowers

I made a posy, while the day ran by:
Here will I smell my remnant out, and tie
  My life within this band.
But Time did beckon to the flowers, and they
By noon most cunningly did steal away,
  And withered in my hand.

GEORGE HERBERT, 1633

Not all posies and bouquets fade and wither away. Some are everlasting. These immortal bunches of flowers make wonderful gifts, which seem to bring the scents and sights of summer to life in the dark winter. Even the smallest garden, as long as it has some of the common herbaceous flowers, can make up enough blooms for an everlasting bouquet.

When dried flowers are mentioned most people think only of the so-called 'everlasting', the straw-flower with its brightly coloured paper-crisp head that is sold in florists' shops. They do not realise that a much wider variety of flowers, foliage and seed heads can be dried to fill up winter vases.

The most charming everlasting gift of this kind is a tussie mussie. These are posies of fragrant herbs which, in times past, were carried to ward off disease. They included bitter herbs whose astringent scent was thought to be particularly prophylactic. Judges were often presented with tussie mussies, in the hope that their scent would keep away the germs of jail fever brought to the courtroom by the prisoners on trial.

Tussie mussies can be given away while still fresh, and kept in water to fade like an ordinary bouquet. But it is more satisfying to preserve them – a simple matter of hanging the posy upside down to dry – so that they will last through the winter, still keeping an echo of their herbal fragrance.

Cut the flowers and foliage from plants such as rue, chives, wormwood, marjoram, thyme, costmary, lavender, sage, balm, mints, camomile, scented geraniums, catmint and other fragrant herbs. Strip the lower leaves. Begin the posy with a single tight rosebud, or a daisy-shaped flower like the elecampane. Surround this with a ring of a bitter-scented, feathery herb such as rue or wormwood. Tie the stems with a piece of wool. Add a second circle of a different flower – chives, for example. Secure with wool again. Continue adding circles of flowers and foliage, arranging them so that the inner rosebud or elecampane is slightly higher than its surrounds.

Make the final circle of firm or evergreen leaves like viburnum, senecio, or whatever your garden supplies. Cut the stems to an equal length and tie the posy with a ribbon, of a colour that matches the dominant colour in the bouquet.

For larger brighter bouquets try sowing some everlastings in your garden. To the French these flowers are not everlasting but *immortelle*, a name which I find more romantic than our own.

The best-known of these is the strawflower, *Helichrysum bracteatum*, an annual which must be started off in the greenhouse or on the windowsill, and then planted outside in the border when all frosts have passed. Ordinary strawflowers grow three to four feet high, but dwarf varieties are also now available.

Other annuals are the greenhouse plant the globe amaranth, *Gomphrena globosa*, and the everlasting sand flower, *Ammobium alatum*, which needs starting in a greenhouse but can be put out in the garden later in the summer. Two others worth trying are the immortelle, *Xeranthemum annuum*, and the Australian everlasting, *Helipterum roseum*.

The sea lavender family, known either as *Statice* or *Limonium*, provides two more annuals – *Limonium sinuatum* and *Limonium suworowii*. Both of these are found in most seed catalogues and both need starting out in a greenhouse or on a windowsill. The family also runs to a useful perennial – the original sea lavender with its rosette of lavender-coloured flowers, *Limonium latifolium*, which can be kept in the flower border throughout its cycle.

DRYING EVERLASTING FLOWERS

Pick the plants a couple of days earlier than the stage at which you want them, for the flowers continue opening after being picked. Strawflowers keep best if picked when in tight buds, and all of them should be picked before they are fully open.

Dry the flowers by hanging them upside down in a dark dry place. Sometimes strawflowers have weak stems and hang their heads after drying. Insert florist's wire into the base of the flowerhead, and use a tiny piece of Sellotape to join it to the stem.

Rather than planting exotic annuals in the border, it is possible to dry those flowers normally found in an English country garden. You will be surprised by the range of colour and beauty available from the average herbaceous border. Most people know of the orange Chinese lantern, *Physalis alkekengi* or *franchettii*, but there are plenty of others too: notably, tansy and the flat yellow flower heads of milfoil, *Achillea filipendulina*, also known as yarrow. The Achillea, says folklore, is so-named because it sprang from the blood of the Greek hero, Achilles, killed besieging the ancient city of Troy. There is a particularly attractive variety of this plant with a silvery tinge to its yellow flowers known as *Achillea taygetea* or 'Moonshine'.

Delphiniums dry well, as do hydrangeas. I pick my hydrangeas at two different times. Picked just before the flowers are unfolding, they dry into a dark purple colour even though the fresh flowers are pink in my garden. Then when the first frosts come, the flower heads turn brick red, a colour which they keep when dried.

Almost all members of the onion family are good candidates – surprisingly, they don't smell. I dry the large Chinese chives, and another ornamental allium whose name I have never discovered but which has large red flower heads. I plan to grow the spectacular *Allium karataviense* soon. Also, wherever I can find a space in my flower border, I cram in any extra leek seedlings I may have over at planting-out time. They shoot up three-foot high seed heads the following year, which dry beautifully.

Other useful common-or-garden plants are corn-flowers, thrift, various lavenders – especially cotton lavender – heather, the larkspurs, sea holly and globe thistles. The latter need picking before they are in full flower otherwise they will shed their seeds

93

later when dry. A plant I plan to grow one day is
Anaphalis which has everlasting daisy-like flowers.

DRYING GARDEN FLOWERS

Pick the flowers on a dry day, two or three days
before they are fully open. Hang them upside
down in a dry dark place.
When they are completely dry, strip the leaves.
If it looks as though the flowers may start
shedding petals, then use a hair spray to help
them stay on.

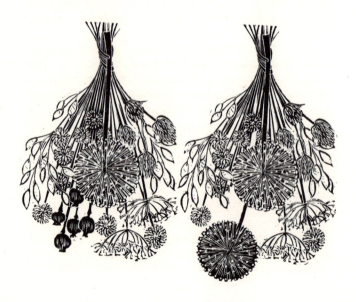

One exception to these rules is the large and fleshy *Sedum spectabile*, sometimes called ice plant, which attracts scores of butterflies with its autumn flower heads. These can be picked when the first tinge of pink appears and will last for two or three months in a dry vase. Do not put dried flowers in the same vase, for its succulent leaves give out water and will turn any companions mouldy.

As well as flowers, it is possible to dry various seed heads. The best-known among these is honesty, *Lunaria annua*, which is simplicity itself to preserve: pick the seed heads when fully ripe and rub off their covering to expose the transparent silvery pennies that remain.

I pick some of my honesty earlier in the year when the seed heads are still green, and then a few weeks later when they have turned a glorious purple colour. I dry these without rubbing off the seed coatings, and at the end of the year I have a mixed bouquet of green, purple and shining silver honesty.

Other possibilities are fennel, opium poppy, oriental poppy, cultivated sorrel, some of the various clematises, including old man's beard from the hedgerow, teasels, some of the irises – notably the stinking iris, *Iris foetidissima*, also known as the roast beef plant because of the scent of its leaves. Its bright berries look wonderful in a vase of dried flowers and foliage.

Grasses to dry include squirrel-tail grass, pampas grass, hare's tail grass, and the trembling heads of pearl grass or quaking grass. To accompany these, pluck one or two heads of barley, wheat or oats, if there are some growing right on the edge of a field where they cannot anyway be harvested.

Pick them before they are fully ripe so that the seeds will not drop out. Hang upside down in a dry dark place. Use hairspray on any heads that, like clematis, have fluff among the seeds. Split the large flowers of pampas grass, using a sharp razor down the stem.

For foliage the best technique is glycerine preserving. Leaves that would otherwise fall off their stems or crumble away, stay flexible for months at a time. The only disadvantage of glycerine is that it turns most foliage dark brown.

All kinds of deciduous-tree leaves can be preserved in glycerine – leaves from the beech, twisted hazel, maple, oak and sycamore. Evergreens like the holly, cypress and other conifers respond well too.

From the shrubs in the garden, use box, mahonia, laurel, cotoneaster, pittosporum, and viburnum. You can try all kinds of interesting foliage, like the foliage on roses, for instance. But one word of warning: variegated leaves often lose their variegation in glycerine.

For the trailing fronds beloved of flower arrangers, preserve honeysuckle, jasmine or even bramble from the hedge. You can glycerine the young berries of elder, tiny immature blackberries and hazelnuts on the stem. Suitable herbaceous-border flowers are Soloman's seal and bells of Ireland and, from the herb garden, try fennel and sorrel.

Two sorts of foliage are outstanding – eucalyptus and the variegated *Elaeagnus pungens variegata*. I have seen preserved eucalyptus foliage which has turned a fine grey-blue, though not possessing a tree

I have not been able to try this for myself.

The new growth of the elaeagnus is an all-yellow leaf. Picked too early in the year, it will simply droop in its glycerine coating. Picked in late August, the foliage turns a bright yellow colour. I would imagine that the two other kinds of yellow elaeagnus – *maculata* and *aurea* would also turn the same colour.

PRESERVING WITH GLYCERINE

Pick the foliage on a dry day. Choose mature leaves, for young ones will droop. But do not try to preserve once they have started to turn colour in the autumn, as they will not draw up the glycerine.

Put one part glycerine to two parts boiling water into a jug. Stir well, then leave to stand till cool. Making a half-inch slit at the base of the larger stems, stand the foliage in the mixture for about 10 to 14 days, until the leaves are supple with a slightly greasy texture. Too long in the mixture, and they will start to ooze glycerine.

The bright yellow foliage of the *Elaeagnus* combines well with dried milfoil, while dark brown holly foliage seems to mix well with dried yellow tansy. A mixture of glycerine-preserved and dried flowers makes a beautiful bouquet.

You can either present the makings of an ever-lasting display, which is perhaps best if the recipient is someone with pronounced tastes of their own, or you can risk handing over an arrangement ready-made – in which case you will need a way of holding it firm. A cheap solution is the ever-versatile lump of Plasticine. Placed at the bottom of a vase, it will anchor the stems; as will dry oasis foam, available from most florists.

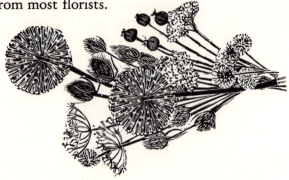

For the art of decorating dried seed heads, I am indebted to Mrs Betty Green, the possessor of a beautiful garden in Cheshire, who passed on to me the secrets of the Christmas decorations she makes every year for charity.

She uses the dried seed heads from opium and oriental poppies, irises, columbines, honesty, and the football-shaped heads of love-in-the-mist.

DECORATING DRIED SEED HEADS

Mix an old bucket half-full of white emulsion paint thinned with water, so that it has the texture of thick milk. Take bunches of the hard seed heads of poppy and iris, dip quickly in the paint and hang upside down to drip dry over old newspaper. Don't immerse for too long otherwise the heads will soften. When they are completely dry spray with gold and white paint.

The same technique can be used with fir, pine and cedar cones. These close up when dipped in the paint, but open up again when put in an airing cupboard on newspaper. The more fragile seed heads can be sprayed directly with gold or silver paint. Always leave a few of the hard heads undipped too because, sprayed directly without the initial white emulsion, they turn out a slightly different colour.

The reason for the emulsion is to provide an undercoat for the expensive spray, which will both save a little and look better.

Betty Green makes lovely table arrangements with the results, placing a candle in the centre surrounded by oasis foam covered in silver foil into which she sticks the decorated seed heads, everlasting flowers and some Christmas baubles.

Christmas always seems to have its last minute rush when you realise you have forgotten to get somebody a present. A very good last minute Christmas present can be made from any winter garden which has evergreens.

AN EVERGREEN BALL

Thread a loop of string, or a piece of wire through a potato in such a way that it can be hung from a hook. Using a skewer poke holes in the potato. Into the holes put sprigs of holly or other evergreens. If you choose sprigs of the same length, you make a regular round ball. I rather like uneven lengths, to give a less formal look. Make sure you put in enough sprigs to hide the potato from all angles. Add a piece of mistletoe to the bottom of the evergreen ball, and you have made a kissing bunch. Christmas baubles, either attached to the sprigs or on the end of a wire poked into the potato, make it more colourful.

The beauty of this simple gift is that it can be made at very short notice from the contents of even a small garden. If you are asked out on Boxing Day at the last minute when the shops are shut and you want to take a gift, you can make an evergreen ball in a few minutes.

Pots
and
Miniature Gardens

And the parson may do much to brighten the gardens of his parish, and so to increase the interest in them by giving plants from his own garden ... I was long ago taught, and have always held, that it is impossible to get or keep a large collection except by constant liberality in giving.

CANON HENRY ELLACOMBE, 1895

A living breathing plant is a beautiful gift. Cut flowers, lovely though they are, will fade and die within days. They bring not just their life but also their death with them into the house. Not so a potted plant – if it is healthy, that is, and well-adapted to life on a windowsill.

There is absolutely no reason why we should not grow and give away our own pot plants. All that is required is patience and the space to grow them. Seed catalogues nowadays offer a wide variety of showy house plants which can be grown from seed, and which in maturity will make a beautiful gift.

POT PLANTS FROM SEED

Try the seeds of busy Lizzies, begonias, calceolarias, ornamental capsicums, castor-oil plants, coleus, cyclamen, umbrella plants, rubber plants, geraniums, polkadot plants, the various primula pot plants, African violets, schizanthus, solanums or winter cherries, sensitive plants, bird-of-paradise flowers and many others. Sow according to packet instructions, and check that you are providing the right degree of heat and adequate light, as well as the right amount of water. Plant out in pots as soon as the seeds have produced two leaves and, as they grow, keep transplanting into larger pots, making sure that the final pot is an attractive one.

Foliage plants like coleus, rubber plant, umbrella plant, polkadot plant and the sensitive plant can be sown at any time. Begonias can also be grown from tubers, which are sold by seedsmen and stocked in most garden shops.

Flowering plants must be timed more carefully so that their flower coincides with the time of giving. Thus cyclamen seeds, which can otherwise be sown any time between August and March, must germinate in August if they're to be ready by Christmas. Many such plants are seasonal and cannot be persuaded to flower out of season simply by altering the date of sowing. And there are plenty which refuse to bloom until they are several years old. The bird-of-paradise flower, for instance, takes three or four years; while the various flowering cacti, which can also be grown from seed, are mostly very slow growers.

Producing pot plants from seed is great fun, but you are likely to get far more than you really want. The local church fete or sale of work will bless you as you hand over the surplus.

Plants like the busy Lizzie, Japanese flowering azaleas, carnations, geraniums and many more can be grown from stem cuttings. Just break off a bit of the plant, preferably at a joint so that you have a heel off a woody stem. In general cuttings are best taken when the plant has finished flowering, unless it is a perpetual flowerer. Some cuttings like azaleas do best in a damp propagator, or in a pot covered with a plastic bag to keep in the moisture. Others, like geraniums, do perfectly well on the windowsill and must not be overwatered. Check the conditions for whatever plant you want to grow. African violets, and most of the succulents will grow from a leaf. Simply place it in compost, and the new flower will grow from its base. I have propagated African violets, *Kalanchoe tomentosa*, the grey-leaved ghost plant, and a wide variety of succulents in this way. Cacti are grown from pups, little buds growing off the main plant.

The problem of taking cuttings, as with sowing seed, is that it requires patience of a high order. The African violet I grew from a leaf did not flower in its first twelve months. After a year my *Kalanchoe tomentosa* was only three inches high and my azalea cuttings, though healthy, still had not flowered.

Plants are programmed genetically. Some may be encouraged to flower with the right light and feeding. Others will persist in waiting for their body clock to tell them when to flower, no matter how good their conditions.

Those like me who grow pot plants on windowsills, rather than in the greenhouse, also have to acknowledge that conditions there are not ideal. The light is not always adequate, and heating usually fluctuates according to human rather than plant needs.

However windowsill nurseries can do a lot more than is generally admitted, so do not be daunted by those who try to insist that a greenhouse with an elaborate heating system is necessary. A year's growth alone on my windowsill has produced some succulents large enough for me to give away as presents.

Perhaps the cheapest way to grow pot plants for Christmas gifts is just to take tender perennials out

of the border in autumn. Busy Lizzie or *Begonia semperflorens* can be potted up before the frosts arrive and will usually continue flowering happily through the winter. After a few weeks they grow into the shape and look of a flower that has spent its life in a pot.

It is also possible to pot up hardy plants like primulas of all kinds and bring them into the house to flower. But remember to do so in plenty of time before they start blooming. If necessary, leave them outside in their pots for a while, bringing them in just before flowering. These outdoor plants will not be very happy in the warmth of a house, so warn their new owners to plant them out again after they have flowered.

The easiest of all to grow for Christmas is the Christmas cactus or, to give it its full title *Schlumbergera x buckleyi*. This is a hybrid raised in 1840 from two different cacti originating in Brazil. Its pink flowers growing out of drooping green segments always seem to arrive by Christmas Day, and it is one of the toughest windowsill plants of them all.

This is such an easy gift to grow that I am surprised more people do not do so. Of course, it takes patience to wait for almost a whole year before your plant is ready to give away. But it would be a good idea, I think, for a child's present – one that could be given to an aunt or family friend the following Christmas.

After flowering, break off pieces which are three or four segments long. Leave for two days for the cut surface to heal. Choose an attractive pot, and place coarse pebbles or clay pot fragments at the bottom with fine compost on top.
Plant the pieces of cactus in a circle round the pot. Keep out of the direct sun until they start growing and water only occasionally, allowing the soil to get completely dry in between times.

If you have only enough segments to plant single ones, they will nevertheless grow to be three segments long by the following Christmas and flower extravagantly.

After years of neglect, plant breeders are beginning to be interested in this humble plant. There are nowadays various different-coloured hybrids on sale. Other succulents suitable for gifts are the Epiphany cactus and the Easter cactus which flower in time for the festivals after which they are named.

One way to make the most out of rather ordinary pot plants is to put them into miniature gardens. It might take too long to grow some of the succulents to a size impressive enough to give away. But while they are still small, they could be planted to make a miniature garden.

First choose your pot, making the best of whatever comes to hand. Plastic can be painted to make it more attractive. Earthenware casserole dishes can be given drainage holes with a mason's drill.

MINIATURE GARDENS IN POTS

Pots without drainage holes need a bottom
layer either of coarse gravel or clay shards to
absorb any excess water. Sterilise these with
boiling water before the soil is added.
The type of compost you use will depend on
the plants. Rather than buy it, you can make
up your own from sterilised peat, loam and
sand (any good gardening book will advise on
proportions). But do not use garden soil, for
this will contain unwanted seeds, spores and
infections.

Finally, use rocks and gravel from the garden as landscaping. Sterilise these too in boiling water before using. Then add the plants. Let the miniature garden settle down for a couple of weeks before giving it away.

I find miniature succulent gardens particularly attractive. They make the ideal gift for the idle windowsill gardener, since they need very little attention. The main dangers are overwatering or too close a proximity to the window on a frosty night.

Cactus expert, Peter Barratt of Sheffield, suggests a mixture of the following for windowsill gardens: aloes, *Euphorbia anaplia, Cleistocactus strausii*, echeverias, mammillarias, *Notocactus warassii*, and crassulas.

If you plan to use pot plants which are not hardy growers, then do not transplant. Bury the pots in the soil of your miniature garden, making sure that only the plant itself can be seen. A nice finishing touch is to include a small jar to hold cut flowers. As before, check carefully that it is completely invisible.

But it's not necessary to have potted plants to hand, or to invest in cacti specially for the purpose. There is plenty of suitable material already growing in the garden: alpines and other small plants from the rockery, for example; not to mention the miniature succulents that grow on most garden walls – the houseleek and various small sedums. If you are prepared to spend a little, though, it is possible to buy spectacularly pretty glass cases nowadays. Those which are entirely closed, like bottle gardens, will need plants specially bought for them. But the ones which have openings to let in the air could be filled with small plants from your garden.

Herbs, for instance. They taste quite different when eaten fresh, and even those who have a garden of their own will welcome the convenience of having their herbs ready to hand – in a window box, for example. Plant those easily grown like chives, thyme, marjoram and winter savory in a long narrow plastic container.

If you are going to include mint, then use a tin opener to open *both* ends of a baked bean tin. Eat the beans, and use the tin to contain the mint. Bury it below the surface of the soil where it cannot be seen, yet where it will prevent the mint roots taking over the whole of the window box.

Another possibility is to use one of those tall pots with holes up the side known as 'tower' pots. Some of these are in rather ugly white plastic which could be painted a less glaring colour. Others are obtainable in terracotta. Starting off with a layer of coarse pebbles or crocks, fill with potting compost. Push the roots of each plant through the holes and firm them in. Move the pot into a shady place for the herbs to take root. An arrangement of different-coloured thymes would look particularly attractive.

Most perennial herbs root from cuttings easily. I have set aside a little bit of my garden as a nursery, and when I am pruning I simply stick three or four cuttings in the earth without more ado. Some of them always take, ready for new plants the following year. And one way of dressing them up for a gift is to make a herb basket.

MAKING A HERB BASKET

Take cuttings of perennial herbs such as sage, winter savory, thyme, rue, rosemary. Take root slips of marjoram, chives and Welsh onion. Grow in identical small pots, whether plastic or clay.

Take a cardboard basket of the kind handed out at those fruit farms where customers help themselves, and cover it with decorated paper or paint it. Fill with the pots, and add a large bow to one side of its handle. You could also consider either covering each pot with the same paper, or painting it the same colour.

Most of these living garden gifts take patience and skill. But the pleasure of growing them is, to my mind, as great as the pleasure of giving them. And as everybody knows, the pleasures of giving are greater than the pleasures of receiving. Best of all, by giving gifts from your garden you will be learning more about gardening.

Vita Sackville-West, one of the most famous of Britain's gardeners, summed it up best. 'The more one gardens, the more one learns; and the more one learns, the more one realizes how little one knows', she wrote. 'I suppose the whole of life is like that: the endless difficulties, the endless fight against one thing or another, whether it be greenfly on the roses or the complexity of personal relationships.'

Packing
and
Prettifying

Nothing so helps a gift as good packaging. Those shops in France where the shop assistant ties up your purchase in wildly expensive wrapping paper with a beautiful ribbon culminating in an extravagant bow, display insight into human nature. Where gifts are concerned, as in so many areas of life, we can be won over by appearances.

Take care and thought about containers. Jam jars, for instance, vary in shape and pattern. Some are plain workaday objects; others, which originally sold the more expensive jam, have pleasing patterns in the glass. Instant coffee jars, with their plastic lids, are particularly useful for pickles and chutneys.

Bottling jars also vary in design. The French unfortunately produce theirs with a kind of metal clamp, but the proportions are pleasant and I usually buy one or two when I am in that country.

English bottling jars vary according to age. The most recent ones, with plastic screw tops, look beautiful in the 2-lb size but far less attractive in the smaller 1-lb version.

Save all pretty jars, no matter whether they come from food or cosmetic products. As long as their contents are adequately washed out and the jar is sterilised, it is fit for use.

In my time I have used squat dark-brown Marmite jars for herbal mustard. I have given away jam in little jars that originally contained hair conditioner,

and I have bottled a few plum olives in a tiny container that once held capers. However, quite often the best choice for your gift is the jar specifically designed for it. Home-made mustard, for example, looks pretty in the little jars which once contained bought mustard.

Many of my jars come from junk or second-hand shops. I have made a habit of looking through those boxes labelled 50p and under and have found some treasures – dimpled whisky bottles, decanters without their tops, a little jar with a frieze of shellfish in the glass.

Corks can be bought from the home-made wine counter of chemists, and whittled down to fit any bottle without its top. Porosan (whose address is in the appendix) make plastic tops to fit 1-lb jam jars – these are especially useful for pickles or chutneys, whose vinegar content would rot a metal top.

Junk shops also offer other treasures. I keep an eye out for old glass biscuit jars. These are occasionally going cheap, and make excellent containers for pickles or for dry potpourri.

Those for whom money is not a problem can buy a wide variety of jars nowadays in ironmongers and department stores. There are blue-green jars in hand-made glass with cork tops, little round jars with black lids suitable for mustards and herbs, and large tall jars which could be filled with pickles. The appendix gives some shopping suggestions.

Take equal care over plant pots. They look much better in clay than plastic. You can also buy them – at a price – made out of china or even complete with decorations. Another trick is to place an unsightly pot inside a handsome one – a teapot which has lost its lid, for a somewhat unusual example.

After you have made your jams and jellies, cover their lids or paper tops with a circle of gingham or floral material. Secure with a rubber band, then cover the band with a ribbon. You will need the circle a little larger than you first think. Do the same with pickles and chutney, choosing a material with a 'savoury' pattern in browns and greens.

When you are giving away a selection of different goodies, match them up – say, a jar of plum gumbo, a jar of plum pickles and a bottle of plum shrub. Then follow that up by packaging them all in the same way – similar labels, and matching cover material and ribbons on all three. This will give a unity to the gift.

Do not hesitate to spend time making a gift look attractive. Jars and boxes and tins can be sprayed with silver or gold paint, or covered with paper if you have deft fingers.

Nowadays all kinds of paper labels and stick-on shapes can be bought. One of my favourites is a little pig. A line of these can march round the pickle jar leading to its identifying label. A stick-on butterfly at an angle adds a cheerful flourish.

It is also a good idea to inscribe your labels with care. Those who study calligraphy know how beautiful the act of handwriting can be. A fountain pen usually improves things, whereas ballpoints can be messy. It may even be worth writing in pencil first, then going over it with a pen. When the ink has dried, the pencil marks can be rubbed off.

Labels can, besides, be vital: at Christmas and other festive times some people receive so many gifts that they are bound to set them aside for later attention – if they have lost your accompanying card, they may wonder whether they have plum shrub or raspberry vinegar. Is the jar full of plum gumbo or three-fruit chutney? And in some cases the name will leave them none the wiser. Add instructions for use wherever these might be necessary.

Indeed careful attention to detail is not just a sign of inner generosity; it can be a pleasure in itself. 'In the very act of giving, I experience my strength, my wealth, my power ...' wrote Erich Fromm in *The Art of Loving*. 'I experience myself as overflowing, spending, alive, hence as joyous.'

Appendix

'The most exquisite delights of the sense are pursued in the contrivance and plantation of gardens; which with fruit, flowers, shades, fountains and the music of birds that frequent such happy places, seem to furnish all the pleasures of the several senses, and with the greatest, or at least the most natural perfections.'

SIR WILLIAM TEMPLE, 1685.

All recipes using fruit and vegetables will vary slightly with the type and condition of what is growing in *your* garden. Test everything by tasting as you go along, and don't be afraid to add or subtract different herbs and spices.

You will need a good supplier of these last from whom to buy orrisroot for the potpourris. Culpeper stock it, together with all kinds of other herbs and spices both dried and in plant form. They also sell very pretty labels. Write for their catalogue to Culpeper Ltd, Hadstock Road, Linton, Cambridge CB1 6NJ.

You can buy attractive labels, plastic tops for jam jars and that vital covering for pickles, Porosan skin, from Thorpac Group p.l.c., Elliot Road, Love Lane Industrial Estate, Cirencester, Glos. GL7 1LU. Boots and John Lewis, as well as many smaller hardware shops, stock Porosan too.

Unusual jars and glass containers can be bought from David Mellor, 26 James Street, London WC2E 8PA, and branches in Chelsea and Manchester.

Stick-on labels in the shape of butterflies, pigs and fairies and a series of pretty little boxes can be bought from Paperchase, 213 Tottenham Court Road, London W1 and branches elsewhere in London and Stirling, Scotland. Many gift and card shops will sell similar goods.

The handiest reference book for people wanting to know more about growing their own house plants is *The House Plant Expert* by Dr D.G. Hessayon, pbi publications.

A wider range of house-plant seeds is available mail order than from the average garden centre. Try Dobies' Seeds, Upper Dee Mills, Llangollen, Clwyd LL20 8SD; Suttons Seeds, Hele Road, Torquay, Devon TQ2 7QJ; and Thompson and Morgan Seeds, London Road, Ipswich, Suffolk IP2 0BA.

The bewildering muddle between the old and the new continues. I have put the equivalent oven temperatures in the text itself, since there is nothing worse than having to halt half way through cooking to find a table converting Fahrenheit to Centigrade. I have chosen to put all weights into the old ounces and pounds, however, since I believe most cooks still work on the old, imperial, system. For those who want to cook in metric, here are the approximate equivalents – they have been either rounded up or down.

Weights

½ oz	10 g	1 lb	450
1	25	1½	700
2	50	2	900
4	110	2½	1 kg 125 g
5	150	3	1 kg 350 g
6	175	4	1 kg 800 g
8	225	5	2 kg 250 g
12	350		

Volumes

5 fl oz	(¼ pint)	150 ml
10	(½)	275
13	(⅔)	360
15	(¾)	425
1 pint	(20 fl oz)	570
3		1 litre 710 ml

Index